Proceedings of the International Workshop on

Applications of Neural Networks to Telecommunications

NEURAL NETWORKS:

The INNS Series of Texts, Monographs, and Proceedings

Series Editor: Harold Szu

ALSPECTOR/GOODMAN/ BROWN, Editors	(1993)	•	Proceedings of the International Workshop on Applications of Neural Networks to Telecommunications
LENDARIS/GROSSBERG/ KOSKO, Meeting Chairs	(1993)	•	World Congress on Neural Networks Proceedings of the 1993 INNS Annual Meeting
LEVINE/ELSBERRY, Editors	(in prep.)	•	Optimality in Biological and Artificial Networks
PRIBRAM, Editor	(in press)	•	Rethinking Neural Networks: Quantum Fields and Biological Data
SOBAJIC, Editor	(1993)	•	Neural Network Computing for the Electric Power Industry

Proceedings of the International Workshop on

Applications of Neural Networks to Telecommunications

Edited by

Joshua Alspector
Bell Communications Research

Rodney Goodman
California Institute of Technology

Timothy X Brown
Bell Communications Research

Psychology Press
Taylor & Francis Group
NEW YORK AND HOVE

First published 1993 by Lawrence Erlbaum Associates, Inc., Publishers

This edition published 2013 by Psychology Press

Psychology Press
Taylor & Francis Group
711 Third Avenue,
New York, NY 10017

Psychology Press
Taylor and Francis Group
27 Church Road
Hove
East Sussex BN3 2FA

First issued in paperback 2014

Psychology Press is an imprint of the Taylor and Francis Group, an informa business

ISBN 13: 978-1-138-88281-2 (pbk)
ISBN 13: 978-0-8058-1560-3 (hbk)

The ability of a competitive entity to utilize information makes it more survivable and adaptive. Communication systems themselves are an organization with their own needs for survivability and subject to economic competition and therefore need adaptive systems. They can take a page from the biology textbook and use neural networks, genetic algorithms and other adaptive and intelligent systems.

The network of networks will need adaptive and intelligent systems to keep it reliable (software reliability), fight off diseases (fault identification), interface with the real world (adaptive user interfaces for information filtering, equalization, speech recognition) respond to threats (congestion control, fraud, network management, rapid rerouting for network failures), keep it efficient, healthy, and growing (resource allocation, market prediction, planning); and these systems must be quick witted, responsive and fast (hardware implementation). The items in parentheses represent work being pursued at telecommunications companies in a highly competitive environment worldwide and they are well represented in the IWANNT program. These companies may be large in size and revenues, have lots of bandwidth and services to sell, and may otherwise be imposing and muscular, but if they are not intelligent, responsive and adaptive to customer needs they will disappear like the large, muscular, but slow-witted dinosaurs of a previous epoch.

This workshop features 50 talks and posters that were selected by an organizing committee of experts in both telecommunications and neural networks. The program deals with the parenthesized topics above and several others such as multiuser detection, character recognition, code design, scheduling, and language identification. This proceedings will also be available on-line in electronic form which will provide multimedia figures, cross-referencing, and annotation. We would like to thank both the INNS and the IEEE Communications Society for their support. On behalf of the organizing committee, we hope you find the workshop a stimulating experience.

Josh Alspector, General Chair

Table of Contents:

PREFACE:

Welcome to the first International Workshop on the Application of Neural Network to Telecommunications (IWANNT). The workshop is motivated by needs in the rapidly changing telecommunications industry which will require new adaptive technologies. We are on the threshold of an explosion in the variety, speed and usefulness of telecommunication networks and neural network researchers can contribute importantly to this emerging new telecommunications infrastructure.

The world is witnessing the rapid evolution of its own nervous system by an unparalleled growth in communications technology. Like the evolution of the nervous systems in animals, this growth is being driven by a survival-of-the-fittest-mechanism. In telecommunications, the entities that fuel this growth are companies and nations who compete with each other. Companies with superior information systems can outrun and outsmart others because they serve their customers better. Similarly, animals with better eyes, ears, brains, and muscles survive in environments where lesser animals fail to.

In the evolution of animals, nature mostly used what it found lying around from previous evolutionary epochs and improved on it. Thus the most primitive part of our brain is the olfactory system on which is built the higher reaches of the cerebellum and cortex. Similarly, the information infrastructure is being built on top of existing POTS, T1 lines, cable systems, using protocols like TCP/IP, ISDN, modems, and using software of various types for network management. To ease the transition into the world of future high speed multi-media information networks, adaptive systems will be needed.

Why is that? Stretching our analogy, humans dominate the planet because of our intelligence and our interfaces to the world and each other. Thus, we see, hear, think, and respond well. We do not see as well as a hawk or hear as well as a dog or are as strong as a horse, but we are more intelligent than any of them, are superbly coordinated for using tools to manipulate our environment, and we can communicate by voice allowing social structures to be built—structures that can further out compete other species to tame a hostile world. The better communication the social structures (families, companies, nations) have, the more competitive they are.

Nowadays we've moved beyond voice as a communications medium -- and courier, mail, books, records, television, magazines, telephones, and videotape as well. Both the quantity of information and the speed of its transmission are increasing at mind-boggling rates. The convergence of all previously mentioned information media into a gigantic digital network of telephone, cable, and wireless networks is nearly at hand.

The ability of a competitive entity to utilize information makes it more survivable and adaptive. Communication systems themselves are an organization with their own needs for survivability and subject to economic competition and therefore need adaptive systems. They can take a page from the biology textbook and use neural networks, genetic algorithms and other adaptive and intelligent systems.

The network of networks will need adaptive and intelligent systems to keep it reliable (software reliability), fight off diseases (fault identification), interface with the real world (adaptive user interfaces for information filtering, equalization, speech recognition) respond to threats (congestion control, fraud, network management, rapid rerouting for network failures), keep it efficient, healthy, and growing (resource allocation, market prediction, planning); and these systems must be quick witted, responsive and fast (hardware implementation). The items in parentheses represent work being pursued at telecommunications companies in a highly competitive environment worldwide and they are well represented in the IWANNT program. These companies may be large in size and revenues, have lots of bandwidth and services to sell, and may otherwise be imposing and muscular, but if they are not intelligent, responsive and adaptive to customer needs they will disappear like the large, muscular, but slow-witted dinosaurs of a previous epoch.

This workshop features 50 talks and posters that were selected by an organizing committee of experts in both telecommunications and neural networks. The program deals with the parenthesized topics above and several others such as multiuser detection, character recognition, code design, scheduling, and language identification. This proceedings will also be available on-line in electronic form which will provide multimedia figures, cross-referencing, and annotation. We would like to thank both the INNS and the IEEE Communications Society for their support. On behalf of the organizing committee, we hope you find the workshop a stimulating experience.

Josh Alspector, General Chair

Programmable VLSI Neural Network Processors for Equalization of Digital Communication Channels

Sa H. Bang, Bing J. Sheu, and Joongho Choi

Department of Electrical Engineering
Powell Hall, Room 604
and Signal and Image Processing Institute
University of Southern California, Los Angeles, CA 90089-0271
sheu@pacific.usc.edu

Abstract

Analog VLSI neural networks are developed for equalization of intersymbol interference and maximum-likelihood sequence estimation. Based on a 4-layered perceptron, the equatizer approximates the optimum receiver. The fabricated chip contains an analog tapped-delay line, (8-12-12-1)-perceptron, and interface circuitry for training. Both simulated and experimental results are presented. A cascadable MLSE receiver using the Hopfield's neural network is also presented.

1. Introduction

Rapid integration of communication, computer, and multi-media technologies has become the key driving force toward global information networking for scientific and engineering applications. Furthermore, the growing demand for physical mobility and service portability is one of the driving market forces for the evolution of personal communications. The performance of digital communication systems over band-limited, multi-path fading channels is largely affected by the ability to overcome channel impairments introduced during signal propagation [1]. The performance of the maximum-likelihood sequence estimation (MLSE) approach is achieved with the exact knowledge of channel characteristics. In addition, the complexity of MLSE approach increases exponentially as the number of channel states increases. The performance of the proposed neural network-based receivers [2] is superior since neither estimation of the channel characteristics nor sampling rate higher than the symbol rate is required. Furthermore, the noise statistics for the proposed receiver needs not to be a white Gaussian.

There have been many researches reported in the area of analog neural network hardware implementations for various applications of adaptive signal processing [3-6]. In our work, the analog VLSI neural network processor chip was designed and fabricated for communication receiver applications. Each programmable synapse cell is made of a compact and wide-range analog multiplier circuit. The arrays of the input neurons and the output neurons are optimized for large current-driving capability. The neural network processor chip performs the feedforward operation and a companion DSP board executes the back-propagation algorithms.

2. Neural-Based Equalizer

2.1 Baseband Model and Neural Network Processors

The neural-based receiver is based on a four-layered perceptron network as shown in Figure 1. In order to accommodate rapid changes in channel characteristics, network training is performed by the extended Kalman filtering (EKF) algorithm to avoid the slow convergence rate of the conventional back-propagation training methods. The channel estimator and the matched-filter are not needed because they are already embedded in the training operations.

By we using the discrete-time model, the received signal is,

$$r(k) = d(k) * h(k) + n(k) = \sum_{i=0}^{L} h(i) d(k-i) + n(k) \tag{2.1}$$

where $d(k)$ is the transmitted signal and $h(k)$ is the channel response. The additive white Gaussian noise $n(k)$ has a zero mean and the variance δ^2. The input to the network becomes

$$y_j = r(k-j)\, y_j = r(k-j) \qquad \text{for}\, j = 1,2,\cdots,M \tag{2.2}$$

The number of input neurons M is to be greater than or equal to the maximum delay-spread of the signal through the channel, normalized by the symbol duration T_S, performing equalization up to L-ray channel models. Since we deal with binary pulse-amplitude modulation (PAM) communication system, the output layer contains only one node.

The numbers of neural units in the input layer, two hidden layers, and the output layer are 8, 12, 12, and 1, respectively. The whole network includes a total of 252 synapse cells, which are fully connected between the corresponding adjacent layers.

A. Programmable Synapses: A compact and enhanced wide-range Gilbert multiplier circuit is used. In order to achieve the wide operation range for the input signal and the synapse weight values, differential pairs are separated. The synapse value is dynamically stored. A refresh cycle of around 0.1 sec is sufficient to retain the 8-bit accuracy [7].

B. Neurons: The neuron of Fig. 2 converts the summed current into the output voltage and generates the sigmoid function. Current-to-voltage conversion occurs through the trans-impedance amplifier consisting of an operational amplifier and a feedback resistor. The virtual-ground scheme for the input current alleviates the problem of the finite output resistance of the synapse cell. The feedback resistor is implemented by six active transistors as shown. Sigmoid function is achieved with the simple inverting amplifiers and the input/feedback resistors, which can change the voltage gain.

C. Input Neuron and SC Analog Delay Line: Since many synapse cells are driven by one input neuron, it is desirable to have a fast settling response for large load capacitance. The input layer is implemented by the switched-capacitor analog delay circuit [9]. The parallel connection in the delay chain avoids accumulation of the offset voltages from the unity-gain amplifiers.

2.2 Simulated and Experimental Results

The operating range for both the input signal and the synapse weight value with the linearity error of less than 1% is -1.75 *V* to +1.75 *V*. The ranges for the positive synapse values and the negative synapse values are well balanced due to the differential-signal scheme. The measured dynamic range of the input neuron is from -3.0 *V* to +3.0 *V*, which is sufficient to drive the synapse cells. The settling time to within 1 % error is about 100 *n*sec for the capacitive load of 2.5 *pF*. The converter can handle the input current level of more than 250 *mA*. The measured voltage gains of sigmoid functions are 6.5, 2.8, and 0.8. Total response time of the neural network processor chip is 0.66 μsec. The die photo of the prototyping processor chip is shown in Figure 4.

Fig. 3 shows a decision boundary of the network for the channel with the minimum-phase transfer function of $H(z) = 0.89443 + 0.44721z^{-1}$ and a signal-to-nose ratio (SNR) of 6 dB. As the number of training symbols increases, the decision boundary closely approximates that of the optimum receiver. The computer simulations and laboratory experiments have been made on the channels with the minimum-phase transfer function and the nonminimum-phase transfer function. Fig. 4 shows comparison of measured (dotted line) and simulated (solid line) results on the convergence performance in the mean-squared-error.

3. Maximum-Likelihood Sequence Estimation (MLSE) using Hopfield's Neural Networks

3.1 Maximum-Likelihood Sequence Estimation

An optimum method of detecting digital data symbols transmitted over time-dispersive, time-varying channels in the presence of additive white Gaussian noise has been known as the maximum-likelihood sequence estimation (MLSE). As a class of nonlinear receivers, it exhibits superior error rate performance compared to its linear counterparts. However, it is often impractical to construct due to the computation intensity and complexity required for the signal processing functions. A more efficient computational method of implementing MLSE is the Viterbi algorithm. An alternative implementation of MLSE function is to use neural networks as a massively parallel computing unit. It can be readily shown that the cost function to be minimized in the MLSE has the same form as the Lyapunov (energy) function associated with the Hopfield's network.

Since the neural network receiver performs the MLSE function once per sequence, the throughput is equal to the network convergence speed times the number of symbols in the sequence. Improvement in the convergence speed increases the throughput. Given a channel memory length, the complexity of the network increases linearly with the length of the sequence. The special structure of MLSE mapping allows a modular architecture for hardware implementation. For the detection of long sequences, multiple receivers can be cascaded to construct a large network.

3.2 Neural Network Implementation of MLSE

The MLSE selects a sequence as the best estimate of transmitted sequence, that maximizes the conditional *a posterior* probabilities or equivalently a sequence that minimizes a quantity

$$J_n = -2\,\mathrm{Re}\left[\sum_k u_k^* y_k\right] + \sum_k \sum_l u_k u_l^* x_{l-k} \tag{3.1}$$

where $\mathbf{u}_n = \{u_1, u_2, \cdots, u_n\}$ is the transmitted sequence of length n, x_k is the autocorrelation function of channel impulse response $h(t)$, and y_k is the sampled output of a matched filter matched to $h^*(-t)$, each sampled at t=kT. The evaluation of values given by (3.1) must be performed over all possible sequences of $\mathbf{u}_n$. Therefore the algorithm complexity is proportional to M^n, where M is the number of signaling alphabets, i.e., $u_k \in \{\mu_1, \mu_2, \cdots, \mu_M\}, \forall k$. In typical data communication in which the length of a sequence is not explicitly given, the number n can be arbitrarily large and in principle could be infinity. If the channel is considered to be time-invariant during n symbol intervals, then $x_k = x_{-k}$, $k = 1,2,\cdots,L$.

A Hopfield [13,14] artificial neural network is proposed for the MLSE function. If the transconductance matrix $\mathbf{W} = [W_{i,j}]$, $i,j = 1,2,\cdots,n$ is symmetric and $g(\cdot)$ is a nondecreasing and differentiable function with respect to the input, then the network is always stable [13]. The network reaches its stable state by minimizing the energy function of the network

$$E = -\frac{1}{2}\sum_{i=1}^{n}\sum_{j=1}^{n} W_{i,j} V_i V_j + \sum_{i=1}^{n} W_i \int_0^{V_i} g_i^{-1}(v)dv + \sum_{i=1}^{n} I_i V_i . \tag{3.2}$$

In general, the state space over which the output $\mathbf{V} = \{V_1, V_2, \cdots, V_{n-1}, V_n\}$ spans is the interior of the n-dimensional hypercube defined by $V_i \in \{0,1\}$. If the gain of neurons at the normal operating point is very high, then the second term in (3.2) vanishes and the minima of E occur only at coners of this space. The similarity between the cost function given in (3.1) and the energy function of Hopfield's network allows a neural network implementation of the MLSE function.

In matrix forms, (3.1) can be written as

$$J_n = -\operatorname{Re}\left[\mathbf{u}^H \mathbf{y}\right] + \frac{1}{2}\mathbf{u}^H \mathbf{X} \mathbf{u} \tag{3.3}$$

where $\mathbf{u} = \begin{bmatrix} u_1 & u_2 & \cdots & u_n \end{bmatrix}^T = \mathbf{u}_I + j\mathbf{u}_Q$, $\mathbf{y} = \begin{bmatrix} y_1 & y_2 & \cdots & y_n \end{bmatrix}^T = \mathbf{y}_I + j\mathbf{y}_Q$, and

$$
\mathbf{X}=\begin{bmatrix} x_0 & x_{-1} & \cdots & x_{-n+2} & x_{-n+1} \\ x_1 & x_0 & \cdots & x_{-n+3} & x_{-n+2} \\ \vdots & \vdots & \ddots & \vdots & \vdots \\ x_{n-2} & x_{n-3} & \cdots & x_0 & x_{-1} \\ x_{n-1} & x_{n-2} & \cdots & x_1 & x_0 \end{bmatrix} = \mathbf{X}_I + j\mathbf{X}_Q .
$$

Since $\mathbf{X}$ is the correlation matrix of $h(t)$, it is Hermitian, i.e., $\mathbf{X}^H = \mathbf{X}_I^T - j\mathbf{X}_Q^T = \mathbf{X} = \mathbf{X}_I + j\mathbf{X}_Q$. In other words, $\mathbf{X}_I^T = \mathbf{X}_I$ is symmetric and $\mathbf{X}_Q^T = -\mathbf{X}_Q$ is skew symmetric. Note that x_0 is real and the main diagonal of $\mathbf{X}_Q$ is all zero. The cost function J_n given by (3.1) is real and needs to be represented by a real quantity in the right hand side for the physical realization of the equation. (3.1) can be expanded and refomulated as

$$
\begin{aligned}
J_n &= -\left[\mathbf{u}_I^T \mathbf{y}_I + \mathbf{u}_Q^T \mathbf{y}_Q\right] + \frac{1}{2}\left[\mathbf{u}_I^T \mathbf{X}_I \mathbf{u}_I + 2\mathbf{u}_Q^T \mathbf{X}_Q \mathbf{u}_Q + \mathbf{u}_Q^T \mathbf{X}_Q \mathbf{u}_Q\right] \\
&= -\left[\mathbf{u}_I^T \middle| \mathbf{u}_Q^T\right]\begin{bmatrix} \mathbf{y}_I \\ \hline \mathbf{y}_Q \end{bmatrix} + \frac{1}{2}\left[\mathbf{u}_I^T \middle| \mathbf{u}_Q^T\right]\begin{bmatrix} \mathbf{X}_I & \mathbf{X}_Q^T \\ \mathbf{X}_Q & \mathbf{X}_I \end{bmatrix}\begin{bmatrix} \mathbf{u}_I \\ \hline \mathbf{u}_Q \end{bmatrix} \equiv -\frac{1}{2}\overline{\mathbf{u}}^T \overline{\mathbf{X}}\,\overline{\mathbf{u}} - \overline{\mathbf{u}}^T \overline{\mathbf{y}} . \qquad (3.4)
\end{aligned}
$$

Note that the matrix $\overline{\mathbf{X}}$ is symmetric. A direct comparision of (3.4) with (3.2) reveals that the desired estimate $\hat{\mathbf{u}}_n$ can be obtained at the output of a Hopfield's neural network if neuron gain is very high,

$$
\mathbf{I} = \overline{\mathbf{y}} = \begin{bmatrix} \mathbf{y}_I \\ \hline \mathbf{y}_Q \end{bmatrix} \text{ and } \mathbf{W} = \overline{\mathbf{X}} = -\begin{bmatrix} \mathbf{X}_I & \mathbf{X}_Q^T \\ \mathbf{X}_Q & \mathbf{X}_I \end{bmatrix}.
$$

In other words, the cost function J_n is the energy associated with the network constructed by the transconductance matrix $\mathbf{W} = \overline{\mathbf{X}}$ and input vector $\mathbf{I} = \overline{\mathbf{y}}$. Figure 5 shows the MLSE receiver realized by Hopfield's neural networks. Each symbol from the matched filter is shifted into a tapped-delay line at each symbol clock and then the neural net begins to find the minimum energy state. To avoid possible local minima, the network is reset to its initial state at the beginning of each operation by applying zero to all inputs. For a given set of samples from the matched-filter, the output of Hopfield's neural network is the best estimate possible. If the channel has the memory of length L, then $x_k = 0$ for $|k| > L$.

For binary modulations, all quantities are real and the cost function to be minimized is reduced to

$$J_n = -\frac{1}{2}\mathbf{u}_I^T\mathbf{X}_I\,\mathbf{u}_I - \mathbf{u}_I^T\mathbf{y}_I = -\frac{1}{2}\mathbf{u}^T\mathbf{X}\mathbf{u} - \mathbf{u}^T\mathbf{y}. \tag{3.5}$$

The matrix $\mathbf{X} = \mathbf{W} = [W_{i,j}], i, j = 1,2,\cdots,n$, is real symmetric and Toeplitz with the element being $W_{i,i+k} = -x_{|k|}, i = 1,2,\cdots,n, |k| = 0,1,\cdots,L$, in the k-th diagonal above or below the main diagonal. The network has n neurons and the $n \times n$ matrix $\mathbf{W}$ has $L(2n - L - 1) + n$ nonzero elements (synapses). The symmetry of $\mathbf{W}$ enables a reduction in the number of synapses by $L(2n - L - 1)/2$. If we define

$$\mathbf{W}_C^T = \left[-\mathbf{W}_U^T \,|\, \mathbf{W}^T \,|\, -\mathbf{W}_L^T\right] = \left[-\mathbf{W}_U^T \,|\, \mathbf{W} \,|\, -\mathbf{W}_L^T\right] \tag{3.6}$$

where

$$\mathbf{W}_U = \begin{bmatrix} x_L & 0 & 0 & \cdots & 0 & 0 & \cdots & 0 \\ x_{L-1} & x_L & 0 & \cdots & 0 & 0 & \cdots & 0 \\ \vdots & \vdots & \vdots & & \vdots & \vdots & & \vdots \\ x_1 & x_2 & x_3 & \cdots & x_L & 0 & \cdots & 0 \end{bmatrix} \text{ and } \mathbf{W}_L = \begin{bmatrix} 0 & \cdots & 0 & x_L & \cdots & x_3 & x_2 & x_1 \\ \vdots & & \vdots & \vdots & & \vdots & \vdots & \vdots \\ 0 & \cdots & 0 & 0 & \cdots & 0 & x_L & x_{L-1} \\ 0 & \cdots & 0 & 0 & \cdots & 0 & 0 & x_L \end{bmatrix},$$

then the Hopfield's network with the transconductance matrix $\mathbf{W}_C$ can be divided into m sub-networks such that $\mathbf{W}_C = \left[\mathbf{W}_0 \,|\, \mathbf{W}_1 \,|\cdots|\, \mathbf{W}_{m-2} \,|\, \mathbf{W}_{m-1}\right]$, where $\mathbf{W}_k, k = 0,1,\cdots,m-1$, is $\mathbf{W}_0$ shifted down by k rows. Furthermore, if $\mathbf{W}$ denotes $\mathbf{W}_0$ with all zero row(s) removed, i.e., $\mathbf{W}_0^T = [\,\boldsymbol{\Omega}^T \,|\, \underline{\mathbf{0}}\,]$, then the original network can be constructed by cascading m subnetworks, each having the transconductance matrix $\mathbf{W}$. The uppermost L and bottommost L feedback inputs associated with $\mathbf{W}_U$ and $\mathbf{W}_L$ are set to zeros, which correspond to the leading and tailing zeros of the sequence to be estimated. In this way, the original network structure is preserved, i.e., $\mathbf{W}_U = \mathbf{W}_L = \underline{\mathbf{0}}$ and $\mathbf{W}_C = \mathbf{W}$. The matched filter matched to channel impulse response $h(-t)$ is approximated by an FIR filter, whose tab coefficients are updated sequence by sequence.

Data communication systems which employs quadra-phase shift keying (QPSK) modulation transmit a sequence of complex-valued data symbols $\{u_k\}$, where $u_k = u_{I,k} + j\,u_{Q,k}$ and $u_{I,k}, u_{Q,k} \in \{-1,+1\}$. (3.4) can be directly mapped onto a Hopfield's network as shown in Fig. 7 for L=2 and n=6. In Figure 6, possible sub-network and modular architecture using multiple sub-networks are also shown. The in-phase and quadra-phase signals from the matched-filter are shifted into two seperate tapped delay lines. During the shift operations, the input switches are opened such that the solution reaches the zero-input response. After n shifts, the switches are closed and the network begins to find its minimum energy state. After a stable state of the network is reached, the estimates of both in-phase and quadra-phase data sequences are available simutaneously at the output. As in binary case, the resulting network for the MLSE can be turned into a modular architecture by adding dummy rows to the transconductance matrix. In this case $\mathbf{W}_U$ and $\mathbf{W}_L$ defined above are required for every sub-matrix of $\mathbf{W} = \overline{\mathbf{X}}$. Fig. 7 shows the resulting network (four identical modules) when L=2 and n=6.

For a typical nonminimum-phase channel ($H(z) = 0.407 + 0.815z^{-1} + 0.407z^{-2}$, $L = 2$), the system equation given in (3.5) is solved. The matched-filter is approximated by a FIR filter whose impulse response is given by $h(-k)$. In Figure 7, several waveforms are depicted during the MLSE of a binary sequence of 38-bit data. Here SNR is set to 15 dB, the gain of neurons is 1000, and neuron input values are exaggerated appropriately. Albeit the received signal is distorted severely due to ISI and noise, the neuron outputs and the corresponding estimates (hard-limited values) are clearly identified.

4. Conclusions

An analog VLSI neural network processor was developed for communication receiver applications. Neither estimation of channel characteristics nor sampling rate higher than the symbol rate is required. Efficient training is performed by the extended Kalman filtering algorithm for inter-symbol interference and white Gaussian noise environments. Measured characteristics of the electrically programmable wide-range synapse cell, the input neuron, and the output neuron are supportive of the precision network operation.

The complexity of the proposed MLSE receiver is proportional to $n \cdot L$, where n is the number of symbols in the sequence to be estimated and L is the channel memory. Therefore, for a given sequence length, the complexity increases linearly with increasing L. Note that the Viterbi algorithm requires arithmetic operations in proportion to 2^L (binary case). Moreover, possible modular architecture allows robust implementations of the MLSE for long sequences. Over many trial runs, it has been shown that the Hopfield's neural networks are good alternative to efficient realization of MLSE function. These are all due to the ability of neural networks to perform optimizations, and a massive parallelism.

References

[1] J. G. Proakis, *Digital Communications*, McGraw Hill, 1983.

[2] S. H. Bang, B. J. Sheu, "A neural-based digital communication receiver for intersymbol interference and white Gaussian noise," *Proc. IEEE Int. Symp. Circuits and Systems*, pp. 2933-2936, 1992.

[3] R. F. Lyon, C. A. Mead, "An analog electronic cochlea," *IEEE Trans. Signal Processing*, pp. 1119-1134, 1988.

[4] C. Koch, et al., *Advances in Neural Information Processing Systems 2*, pp. 750-757, Morgan Kaufmann, 1990.

[5] A. Moore, et al., "A real-time neural system for color constancy," *IEEE Trans. Neural Networks*, pp. 237-247, 1991.

[6] J. Van der Spiegel, et al., " An analog neural computer with modular architecture for real-time dynamic computations," *IEEE Jour. Solid-State Circuits*, pp. 82-92, 1992.

[7] B. W. Lee, Bing. J. Sheu, *Hardware Annealing in Analog VLSI Neurocomputing*, Kluwer Academic, 1991.

[8] M. Banu, Y. Tsividis, " Floating voltage-controlled resistors in CMOS technology," *Electronic Letters*, pp. 678-679, 1982.

[9] Y. S. Lee, K. W. Martin, " A switched-capacitor realization of multiple FIR filters on single chip," *IEEE Jour. Solid-State Circuits*, pp. 536-542, 1988.

[10] G. J. Gibson, et al., "The application of nonlinear architectures to adaptive channel equalization," *Proc. IEEE Int. Conf. Communications*, pp. 649-653, 1990.

[11] J. Provence, "Neural network implementation for an adaptive maximum-likelihood receiver," *Proc. IEEE Int. Symp. on Circuits and Systems*, pp. 2381-2385, 1988

[12] S. Singhal and L. Wu, "Training multilayer perceptrons with the extended Kalman algorithm," in D.S. Touretzky [Ed.], *Advances in Neural Information Processing Systems, Volume I*, Morgan Kaufmann, New York, 1990

[13] J.J. Hopfield, "Neurons with graded response have collective computational properties like those of two-state neurons," *Proc. Natl. Acad. Sci. U.S.A.*, Vol. 81, pp. 3088-3092, 1984

[14] D.W. Tank and J.J. Hopfield, "Simple 'neural' optimization networks: An A/D converter, signal decision circuit, and a linear programming circuit," *IEEE Trans. on Circuits and Systems*, Vol. 33, no. 5, pp. 533-541, May 1986

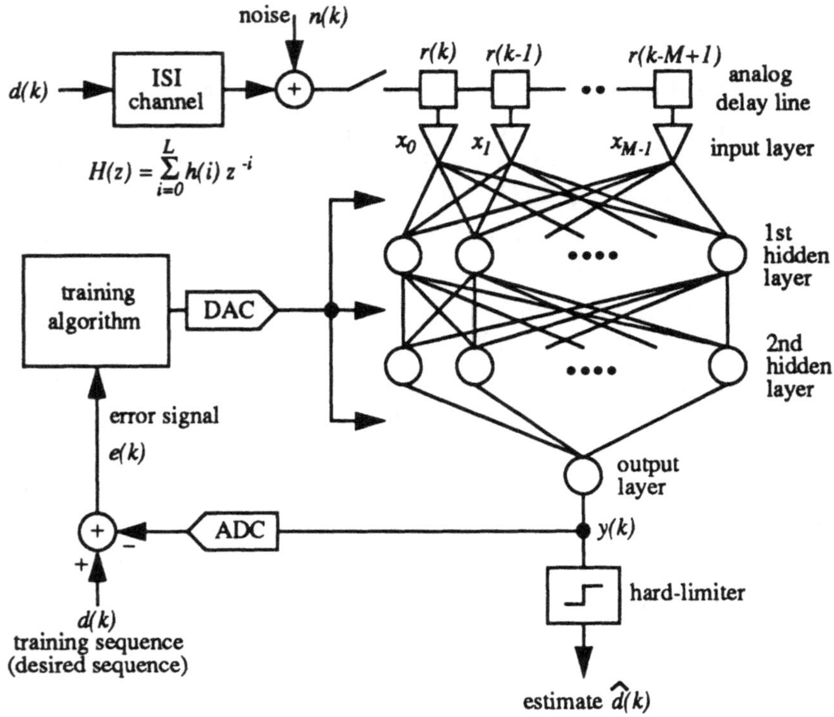

Figure 1. Neural-Based Communication Receiver

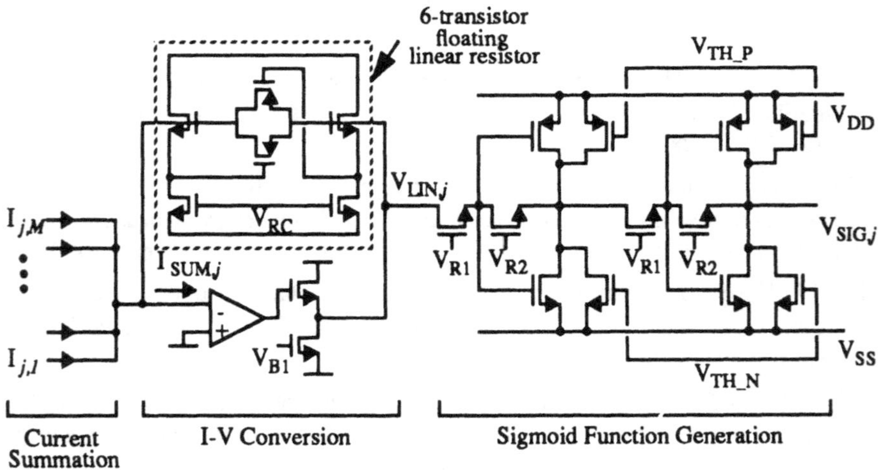

Figure 2. Circuit Schematic of Neurons (Sigmoid Function Generator)

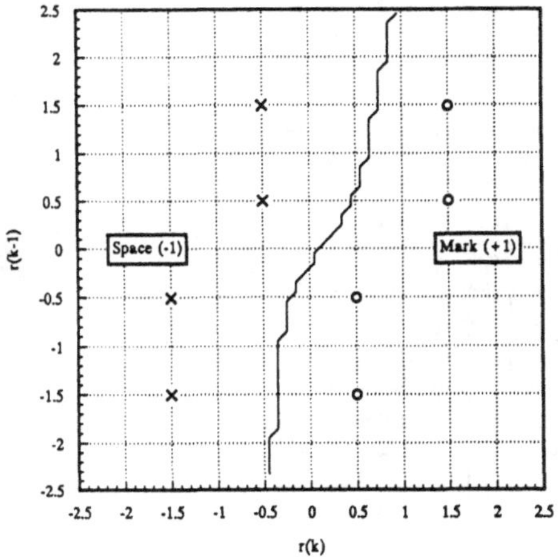

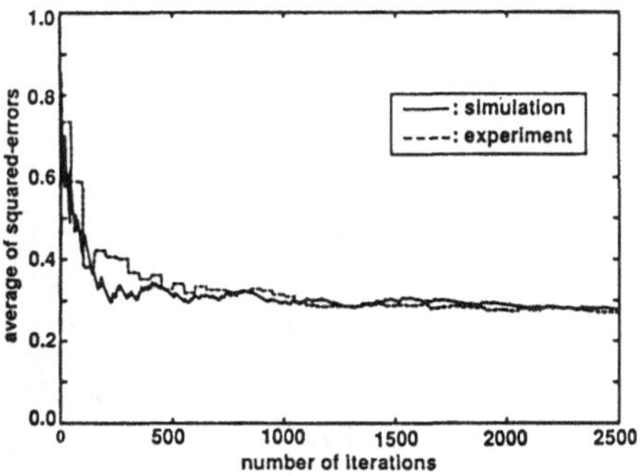

Figure 3. Decision Boundary for Minimum Phase Channel $H(z)$=0.89443+0.44721z^{-1} (-1 in the left side and +1 in the right side)

Figure 4. Comparision of Simulated and Measured Results $H(z)$=0.89443+0.44721z^{-1} and SNR=6 dB

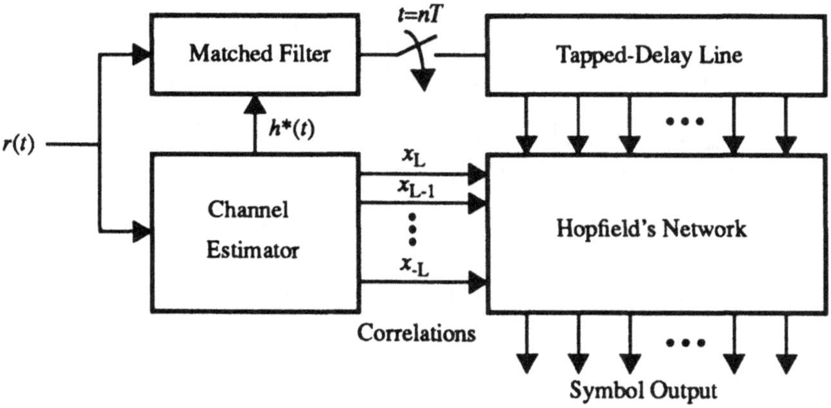

Figure 5. Optimum MLSE Receiver using Hopfield Neural Network

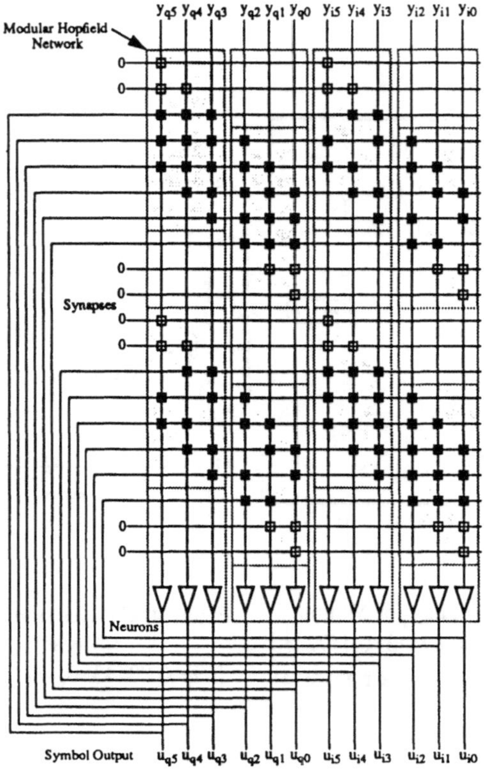

Figure 6. MLSE Neural Network (Modular) for QPSK Modulation (L=2, n=6)

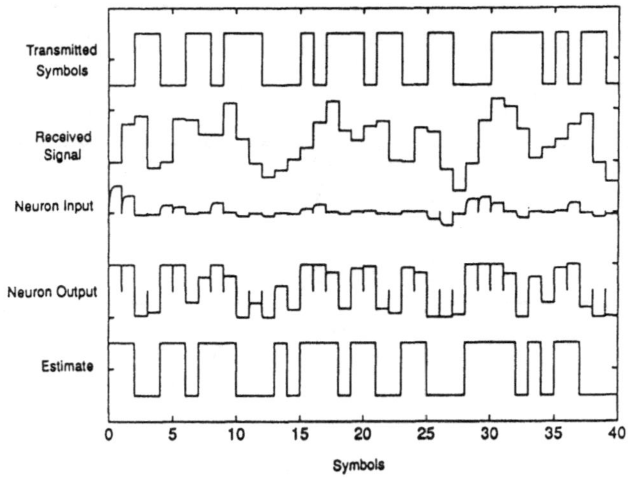

Figure 7. MLSE of 38-Bit Data (n=38, L=2, $H(z)$=0.407+0.815z^{-1}+0.407z^{-2}, SNR=15 dB)

An Analog Neural-Network Co-processor System for Rapid Prototyping of Telecommunications Applications

Anthony Jayakumar and Joshua Alspector

Bellcore
445 South Street, Morristown, New Jersey 07960-6438 USA
tony@bellcore.bellcore.com
josh@bellcore.bellcore.com

Abstract

We describe an experimental neural network learning system that is designed to rapidly prototype telecommunications applications. It functions as a co-processor in a Sun4 workstation through the VME bus interface and is controlled by a friendly, graphical user interface. The experimental prototype system uses the cascadable analog learning chips that have been described in a previous publication [1]. A four-chip network consisting of 128 neurons and about 2000 adaptive synapses has been successfully tested. It can be used for 1) classification problems by use of feedback or feed-forward neural networks, 2) optimization problems by use of mean-field and Boltzmann settling, or 3) problems that require high speed learning. As an example of a problem that requires continuous, high speed adaptation, the system has been used to prototype a simple channel equalization problem. The experimental system is functioning as a general purpose test bed for running neural network applications.

1 Introduction

Neural network chips have reached a stage of maturity sufficient to solve sophisticated, real-time, and real-world problems. There are now several neural network hardware systems that are being offered for sale by various suppliers. Several of these devices are high-performance, high-power digital computing platforms. There are also many application domains where low-power, low-cost, adaptive, analog processing is best. In the present stage of development, these analog systems can only realize their full potential if they interface with ubiquitous digital computing hardware and user friendly software to enable rapid development and deployment of embedded neural solutions for demanding applications.

We describe a neural network learning system that is designed to fulfill the above need. It functions as a co-processor in a Sun4 workstation through the VME bus interface and is controlled by a friendly, graphical user interface. The system uses the cascadable analog learning chips that were described at NIPS 4 [1]. A four-chip network consisting of 128 neurons and about 2000 adaptive synapses has been successfully tested. The system is functioning as a general purpose test bed for running neural network applications.

2 The Learning Chip

The neural network chip that is used in the system is an analog learning device that uses the Boltzmann Machine (BM) and the Mean Field (MF) learning algorithms for on-chip weight updates[1]. The chip contains 32 neurons and 496 adaptive synapses. The network dynamics are fully analog but with a 5-bit digital weight storage and update. Gradient descent is performed through stochastic and/or deterministic simulated annealing. An on-chip 32-channel uncorrelated noise generator is the 'heat source' for the noise annealing. The injected noise and the gain of the neuron can be varied for the BM and MF settling procedures respectively. In addition, to permit easier cascadable connection of neuron chips, a chip containing 1024 adaptive synapses has also been fabricated.

The performance of the chip is rated at about 10^8 connection-updates per second. Roughly 100,000, 32-bit digital patterns/second or 32 analog channels, band-limited to 50 KHz each, could be fed into the chip for fast learning or classification. The performance figures in a cascaded, multi-chip network would rise linearly with the number of chips used.

3 The Co-processor System

To preserve the high throughput to the neural network and to maintain ease of use through a graphical user interface software environment, the learning system was envisioned to function as a co-processor to a workstation. This enables the validation of neural network methods to a specific application before any custom hardware is built. The system would also serve as a test bed for researchers to try out their applications locally or remotely on the hardware through an X-windows based graphical interface.

The system consists of the following components: 1) Controller board, 2) Analog and digital pattern generation/acquisition boards and 3) Neural network boards. These boards constitute a VME slave sub-system that can be easily interfaced with any high-performance computing platform. A simplified diagram of the co-processor is shown in Figure 1.

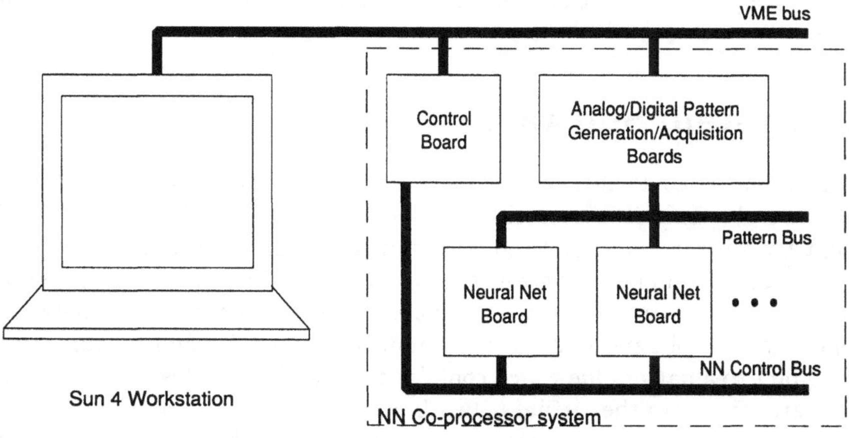

Figure 1: Neural Network Co-processor System

3.1 The Controller board:

This board handles the VME interface between the neural network and the host. Address translation circuitry translates higher level commands into control signals for the neural network (NN) chips. The control functions performed by this board are weight read/writes, neural network topology setup, noise and gain annealing schedule generation for relaxation of the network, learning control and seeding the noise generator. This controller can support up to 16 neural network chips.

3.2 The Pattern Generation/Acquisition system:

This system consists of a digital board for fast 64-bit digital input pattern presentation and 32-bit readouts from the NN chips. Off-the-shelf, fast multi-channel A/D and D/A conversion boards enable 32-channel analog pattern presentation/acquisition to and from the neural network. The analog I/O signals are routed to the NN boards through a private analog bus which serves as the main 'pattern' bus for the network.

3.3 The Neural Network board:

This board houses the neural network chips. Up to six chips can be cascaded in one board and any number of NN boards can be used limited only by connectivity constraints. This board is shielded from the VME bus to reduce the impact of high speed switching noise and is connected to the host through the controller and pattern cards via the control and pattern buses.

4 System Software

The system is controlled by our in-house neural network simulation tool, 'Braincore'. This is a versatile, X-windows based, graphics oriented, C language software package designed to simulate various NN algorithms. The basic graphics interface controls the operation of the hardware through a series of Unix system calls to the co-processor. The user can set up a network topology, control parameters such as annealing schedules and neuron gains and point to a pattern file which contains the training and test vectors. The software translates the virtual network map to physical connections on the cascaded chip network, downloads the annealing schedules to the on-board memory, converts them to analog voltages, puts the patterns on the pattern bus and activates the learning cycle. To the user, the hardware is totally transparent and just looks like another simulation module to be invoked, except now it runs on the fast co-processor hardware.

5 Experimental Results

The system has been designed for general purpose use by researchers to aid the speedy deployment of neural network techniques in real-time hardware after extensive study and simulation of the problem by conventional neural network methods but before any custom hardware has to be built. Currently we are interested in the problem of fast adaptive equalization of highly non-linear channels, such as wireless channels encountered in Personal Communications Systems. The co-processor hardware can validate the implementation schemes in a fairly realistic environment.

Fig. 2 shows the X-windows display of Braincore running the equalization task for a noisy channel but with the learning done in our hardware system. The software presents bit patterns to the hardware and displays the results. The neural network structure is shown along with Hinton diagrams for the tap weights on the equalizer. The two graphs on the display are learning curves for equalization of a channel with the transfer function

$$H(z) = 0.3482 + 0.8704z^{-1} + 0.3482z^{-2} \tag{1}$$

with a signal-to-noise ratio of 20 dB. The top plot is the percentage of correct decisions while the lower curve is the mean squared error. One can see that the network approaches the performance of an optimum receiver after a few thousand training examples.

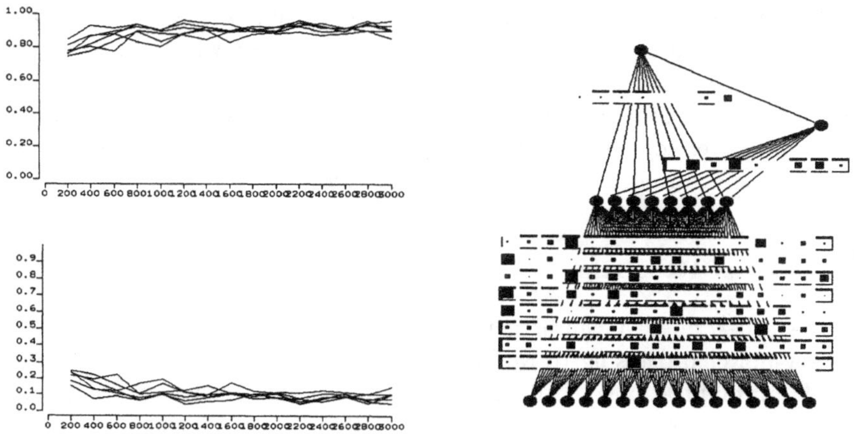

Figure 2: Learning in neural network equalizer

6 Other Applications

We expect to be able to prototype many other telecommunications applications on this system because of its many modes of operation. These include 1) classification (either in feedback or in feed-forward), 2) optimization using the settling mechanisms of noise and gain control, and 3) learning for real-time adaptation.

For classification tasks, the weights can be learned by the actual system beforehand and used in the feedback mode. The system would classify by settling using gain or noise variation which requires 10-50 μs per pattern. This should be sufficient for most classification applications. It may, however, be desirable to use a feed-forward system for classification in those cases where the weights have been learned by an off-line technique such as back-propagation. In that case, the system has the capability of downloading fixed weights from a file and removing one direction of its bi-directional synapse from operation. The system will not need to settle its feedback paths and will be faster (about 3 μs per pattern).

For optimization tasks, one can use the built-in settling features of the feedback network with downloaded weights appropriate for the task. This could be used for switching and routing problems using the techniques of Brown [2] by prototyping winner-take-all circuits with inhibitory feedback connections. One could also use the Hopfield style optimization technique for crossbar switching as advocated by Marrakchi and Troudet [3]. Other problems where these techniques can be applied include dynamic bandwidth allocation in wireless networks [4], and packet routing [5].

For real-time adaptation, this system uses its capabilities fully. We have already described an application in channel equalization which is best suited for noisy or non-linear channels [6]. We are currently exploring the use of this technique in other demanding telecommunication channels such as high-speed data over copper (HDSL,ADSL). This technique is also applicable to noise or echo-cancellation filtering in these channels.

7 Conclusions

We have described a neural network co-processor system that uses analog low-power processing to rapidly prototype many potential applications in telecommunications. This system would aid researchers to design and deploy neural network hardware based solution to many demanding applications.

Acknowledgements

This research was supported by the Advanced Research Projects Agency of the Department of Defense and was monitored by the Air Force Office of Scientific Research under contracts F49620-90-C-0042 and F49620-92-C-0075. The United States Government is authorized to reproduce and distribute reprints for governmental purposes notwithstanding any copyright notation hereon.

References

[1] J. Alspector, and A. Jayakumar and S. Luna, "Experimental Evaluation of Learning in a Neural Microsystem," in *Proceedings of Neural Information Processing Systems 1991,* vol. 4, pp. 871–878.

[2] T.X. Brown, " Neural Networks for Switching," in *IEEE Communications Magazine,* vol. 27, no. 11, pp. 72–81, November 1989.

[3] A.M. Marrakchi, and T. Troudet, "A Neural Net Arbitrator for Large Crossbar Packet-Switches," in *IEEE Transactions on Circuits and Systems,* vol. 36, no. 7, pp. 1039–1041, July 1989.

[4] D.Kunz, "Channel Assignment for Cellular Radio Using Neural Networks," in *IEEE Trans. on Vehicular Technology,* vol. 40, no. 1, pp. 188–193, Feb. 1991.

[5] M.M. Ali and H.T. Nguyen, "A Neural Network Implementation of an Input Access Scheme in a High Speed Packet Switch," in *Proceedings of the 1989 Global Telecommunications Conference,* pp. 1192–1196.

[6] S. Chen and G.J. Gibson and C.F.N. Cowan, "Adaptive channel equalisation using a polynomial-perceptron structure," in *IEE Proceedings,* vol. 137, pp. 257–264, October 1990.

Improving Conventional Equalizers with Neural Networks

Jesús Cid-Sueiro, Aníbal R. Figueiras-Vidal*

ETSI Telecomunicación-UV, Valladolid, 47011 Spain
cid@gtts.ssr.upm.es
* *DSSR, ETSI Telecomunicación-UPM, Madrid, 28040 Spain*
weruaga@gtts.ssr.upm.es

Abstract

In this paper we show that the low detection capabilities of conventional equalizers (linear and decision feedback equalizers) and the excessive complexity of those based on neural networks can be avoided by means of mixed schemes. Linear equalizers aided by centroid selection (as in Radial Basis Function networks) improve performance over the standard linear FIR equalization approach, and modified DFE based on bi-layer perceptron avoids error propagation, outperforming conventional schemes.[1]

1 Introduction

This paper deals with the problem of symbol detection in digital communication systems, starting from the digital channel model of Figure 1. The transmitted sequence x_k is affected by a possibly non-linear distortion and an additive white noise n_k. The detector (equalizer) tries to restore the transmitted sequence by using the observation r_k, with the goal of minimizing the expected number of errors.

Symbol-by-symbol digital channel equalizers based on neural networks have been recently proposed (see [1] [2] [3], for example) in order to solve the

[1]This paper has been partially supported by CICYT grant TIC 92 # 0800-C05-01

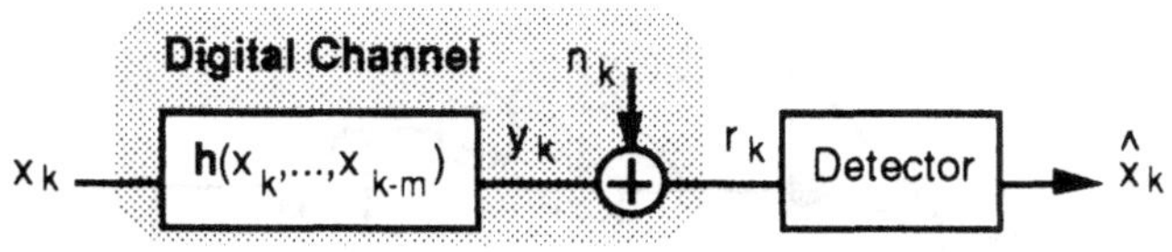

Figure 1: Digital Channel Model

low detection capabilities of classical schemes (linear equalizers, or decision feedback equalizers (DFE)). Usually, these novel schemes reach smaller error probabilities than that of linear structures, at the expense of very long training times and a complexity excessively growing with the size of the channel or the equalizer memory (i.e. the number of received samples used in a single decision).

Another approaches [4] have pointed out the capability of Radial Basis Function (RBF) networks to compute optimal symbol-by-symbol decisions. RBF equalizers based on this feature can be easiliy trained (via learning the channel impulse response and the noise statistics), since there is a direct relationship between the channel parameters and the coefficients of the neural network. Furthermore, it is shown in that paper that recurrent RBF networks (RRBF, in the following) like that of Figure 2 can compute the optimal symbol-by-symbol decisions using all the received samples. Using RBF based schemes, the learning problem is solved, but the complexity is exponentially growing with the length of the channel impulse response. The purpose of this paper is to show that combining classical and neural structures, the performance of conventional equalizers can be improved, avoiding the complexity of neural schemes.

2 Simplified RBF equalizers

In [4] we also propose several methods in order to simplify RRBF equalizers. One of them is based on making decisions about past samples. We analize here the network resulting from feeding back x_{k-2}, which is shown in Figure 3. The structure is similar to that of Figure 2 with an aditional feedbak removing the post-cursor intersymbol interference (ISI) of delay longer than 1.

Assumming that the long-delay ISI is perfectly removed, the feedforward part of Figure 3 is the optimal symbol-by-symbol detector, and the diferent computations in the network have the statistical meaning shown in Figure 2. The decisions are made by maximizing $p_i = Pr\{x_k = i \mid r_k, r_{k-1}, \ldots\}$, being

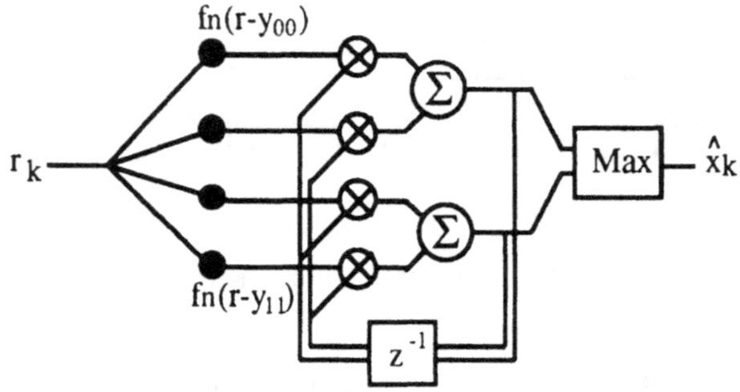

Figure 2: Recurrent RFB equalizer for a channel with memory 1 and a binary transmission

i equal to +1 or -1 in a bipolar binary transmission. Then, by noting with R_k the vector of samples $(r_k, r_{k-1}, \ldots)$ we can write

$$\begin{aligned} p_1 &= Pr\{x_k = 1 \mid R_k\} = \\ &\quad Pr\{x_k = 1 \mid x_{k-1} = 1, R_k\} Pr\{x_{k-1} = 1 \mid R_k\} + \\ &\quad Pr\{x_k = 1 \mid x_{k-1} = -1, R_k\} Pr\{x_{k-1} = -1 \mid R_k\} \end{aligned} \tag{1}$$

Defining $q_1 = Pr\{x_{k-1} = 1 \mid R_k\}$, then

$$\begin{aligned} p_1 &= q_1 \frac{f_n(r_k - (h_0 + h_1))}{f_n(r_k - (h_0 + h_1)) + f_n(r_k - (-h_0 + h_1))} + \\ &\quad (1 - q_1) \frac{f_n(r_k - (h_0 - h_1))}{f_n(r_k - (h_0 - h_1)) + f_n(r_k - (-h_0 - h_1))} \end{aligned} \tag{2}$$

where f_n is the Gaussian noise pdf. It is not hard to see that (2) is equivalent to

$$p_1 = q_1 sigm(\frac{2h_0}{\sigma^2}(r_k - h_1)) + (1 - q_1) sigm(\frac{2h_0}{\sigma^2}(r_k + h_1)) \tag{3}$$

where $sigm(.)$ is the sigmoidal function. The coefficient q_1 can not be easily computed, but if we assumme that there is no relevant information about $x_k - 1$ in r_k, then we can approximate q_1 by

$$q_1 \approx Pr\{x_{k-1} = 1 \mid R_k - 1\} \tag{4}$$

that has been computed by the RRBF network in the past time instant. The computations in (3) and (4) can be realized in a recurrent bi-layer

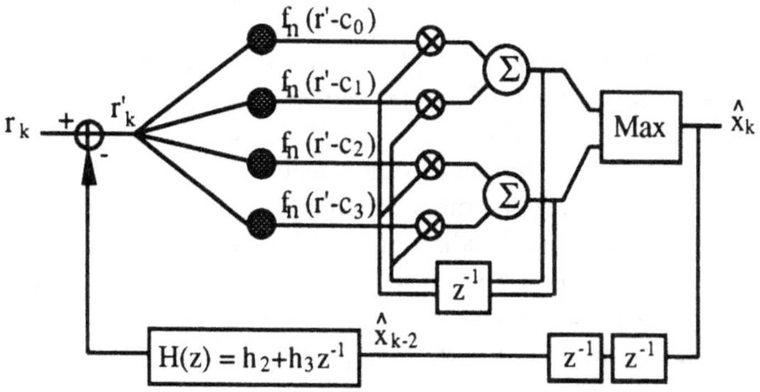

Figure 3: Simplified RRFB equalizer for a linear channel $H(z) = h_0 + h_1 z^{-1} + h_2 z^{-2} + h_3 z^{-3}$

perceptron (BLP) structure like that of Figure 4. Training can be carried out by learning the channel parameters (impulse response and noise variance) and then computing the network coefficients in a direct way. The long training times of backpropagation and other algorithms are thus avoided, and the number of sigmoids is not dependent on the size of the channel, avoiding complexity problems. Even, in spite of the aproximation of (4), some useful statistic information about decisions is feedback, reducing the error propagation effects of classical DFE schemes.

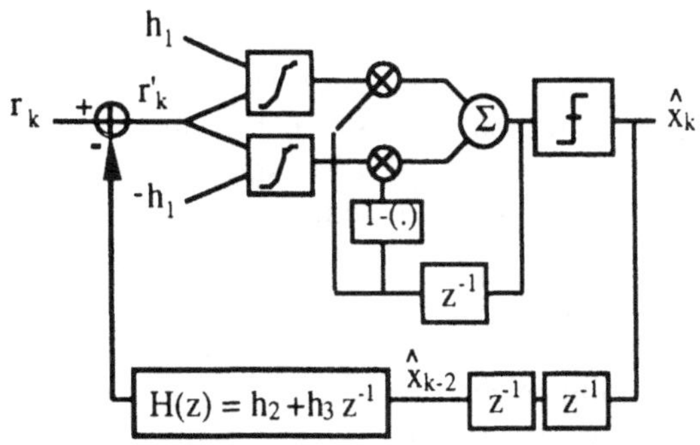

Figure 4: Bi-layer perceptron equalizer for a linear channel $H(z) = h_0 + h_1 z^{-1} + h_2 z^{-2} + h_3 z^{-3}$

3 Piecewise linear equalizers

A similar simplifying method can be used to reduce complexity in RBF equalizers. Assume that long-delay ISI removed by means of feedbak. Then we can use an RBF network to make decisions based in the last two received samples, as in [1] [4] [5], but including a feedbak loop of like in Figure 4. In these papers, the finite-memory equalization is viewed as a geometric classification problem: the equalizer traces decision boundaries partitioning the 2-dimensional space of the received samples in regions, and each one of them is assigned to one symbol of the transmitted alphabet. In the 2-dimensional case, these boundaries are simple and can be aproximated by piecewise linear curves.

The optimal boundary is directly related to the position of the centroids, i.e. the vectors of all the possible received samples without noise [1], and they can be computed from the channel impulse response, so they can be easily learned. Besides, when the channel response is linear and the ISI part with delay longer than 1 or 2 is removed, the pieceweise linear boundaries can be estimated using just the most external centroids. Following this idea, we can construct a simple piecewise linear equalizer (PLE), like a DFE structure in wich the feedbak just removes the long-dealy ISI, and the simple hard limiter in the feedforward part is replaced by a piecewise linear system. As before, fast training can be carried out by learning the channel response.

4 Simulation results

We have compared the proposed schemes with the optimal symbol-by-symbol equalizers, in order to measure the performance loss caused by the symplifying assumptions involved in the derivation of the BLP and the piece-wise linear equalizers. We consider the linear channel $H(z) = 1 + 0.5z^{-1} + 0.25z{-}2$ corrupted by additive white Gaussian noise. Figure 5 shows the symbol error probabilities of the proposed schemes for different values of the signal-to-noise ratio.

In this example, and in all the simulations we have carried out, the symbol error probabilities reached by the BLP equalizer are nearly equal to that of simplified RRBF equalizers. The performance loss of the PLE is small, but significant. The structure is simple and easy to train, but there is a penalty in performance because there is no feedback of statistics in the structure. Even, in some cases, simpler DFE structures worked better.

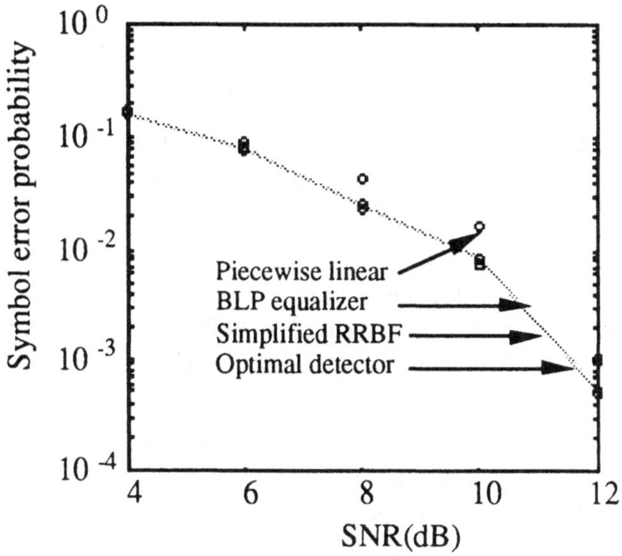

Figure 5: Comparison between equalizers for the linear channel $H(z) = 1 + 0.5z^{-1} + 0.25z^{-2} + 0.1z^{-3}, and different values of SNR$

5 Conclusions

In this paper we show several possibilities of combining conventional equalizers with neural structures, with the goal of improving performance avoiding the excessive complexity problems of other schemes. The proposed networks can be trained via learning the channel parameters (impulse response and noise variance), and then computing network parameters in a direct way. Thus, training is fast. Furthermore, some of the proposed networks (symplified RRBF equalizers and BLP equalizers) provide aproximated statistics about decisions, and they can be used for updating the channel estimate in a non-supervised and non-decision directed way, following similar ideas to that proposed in [7]. Our future research will follow this line.

References

[1] G. J. Gibson, S. Siu, C.F.N. Cowan, "The Application of Nonlinear Structures to the Reconstruction of Binary Signals," *IEEE Trans. Signal Processing,* vol. 39, no. 8, pp. 1887-1884, 1991.

[2] S. Arcens, J. Cid-Sueiro, A. R. Figueiras-Vidal, "Pao Networks for Data Transmission Equalization," in *Proc. of the International Joint Conference on Neural Networks,* vol. 2, Jun. 1992, pp. 963-967.

[3] T. Kohonen et al, "Combining Linear Equalization and Self-Organizing Adaptation in Dinamic Discrete-Signal Detection," in *Proc. of the International Joint Conference on Neural Networks,* vol. 1, Jun. 1990, pp. 223-228.

[4] J. Cid-Sueiro, A. R. Figueiras-Vidal, "Recurrent Radial Basis Function Networks For Optimal Symbol-By-Symbol Equalization," presented at the 3rd Cost #229 Workshop on Adaptive Algorithms in Communications, Bayona, Spain, June 1993.

[5] C.P. Callender, C.F.N. Cowan, "'Two Novel Non-Linear Approaches to Channel Equalisation for Digital Communications," in *Proceedings of the 2nd Cost #229 Workshop on Adaptive Algorithms in Communications*, Bordeaux, France, Oct. 1992, pp. 247-254.

[6] J. Cid-Sueiro, A.R. Figueiras-Vidal, "Optimal Symbol-by-Symbol and Sequential Nonlinear Channel Equalizers Using Neural Network Structures," in *Proceeding of the First Cost #229 WG. 4 Workshop on Massivelly Parallel Computing*, Leysin, Switzerland, March 1992.

[7] J. Cid-Sueiro, A.R. Figueiras-Vidal, "Optimal Blind Equalization of Gaussian Channels," in *Proceedings of the International Workshop on Artificial Neural Networks, IWANN'93*, Sitges, Spain, June 1993.

Neural Networks for Adaptive Equalization

Timothy X Brown

Bell Communications Research
445 South Street, Morristown, NJ 07960
timxb@faline.bellcore.com

Abstract

Deficiencies in linear approaches to equalization have led to non-linear neural network equalization. Because of the potential for efficient analog hardware implementation, neural networks are especially attractive. A review of the neural literature reveals that indeed neural networks can have improved performance, but at the expense of much longer training times. In this paper we present the results of this review as well as methods for counteracting the recognized deficiencies.

1 Introduction

Within a communications system, equalization is the task of compensating for distorted received data so as to minimize symbol detection errors. Of special interest is the case when the channel is time varying and the equalizer must track these changes. A good summary of the field is given in [13]. The first successful approaches to this problem were wrought 30 years ago [12], and they still are the basis of most modern approaches. These, though, are linear approaches in that they are simple linear thresholds on periodic samples of the received signal. Some non-linearity is introduced by also including in the sum past threshold outputs. The net architecture is shown in Figure 1.

Some researchers have argued that there are deficiencies in this model, and have suggested that more complicated functions than simple linear threshold

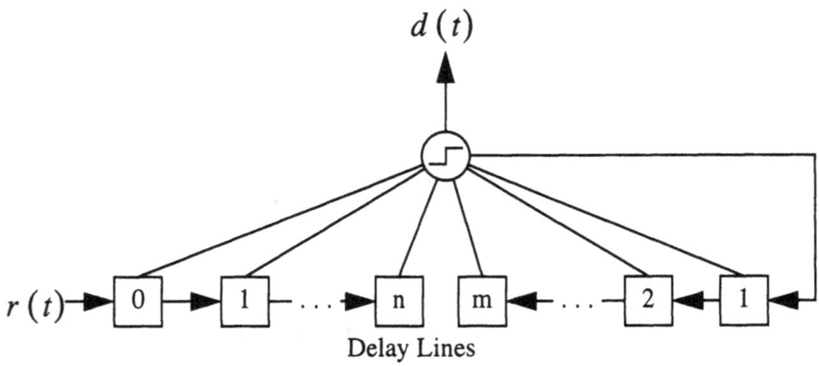

Figure 1: Equalizer structure showing inputs that are both feed-forward on received symbols and feed-back on detected outputs.

functions should be used such as Volterra series (weighted sums of first, second, etc. order polynomials) on the taps[5]. Another obvious approach is to use neural networks. This paper motivates why neural networks are necessary, reviews published neural approaches, and gives concrete recommendations for combatting deficiencies recognized in these approaches.

2 Why Neural Networks?

The argument for neural networks hinges on two basic points. One is that as more complex channels are being pushed to carry increasing volumes of data, assumptions of linear channel responses become less tenable. Even simple linear models under Gaussian noise produce non-linear optimal decision surfaces[5]. A second reason is that the analog implementations of neural networks have the potential for low complexity (i.e. VLSI area) and low power—suitable for the profusion of portable applications—while at the same time providing the necessary processing speed. High-speed digital equalizers [9] can use nearly a Watt not including the power used by the necessary analog to digital convertors (ADC). Analog neural hardware even with on-chip learning consumes considerably less [2]. Avoiding an expensive ADC step garners significant additional savings since the analog signals can be fed directly to the neural network via a low-power, high-speed charged couple device (CCD) or switched capacitor based tap delay line. Calculations based on actual VLSI circuit models predicts a factor of 20–100 lower power consumption over similar size and speed digital equalizers.

3 What Has Been Shown Already

Neural network approaches to equalization have been simulated by a variety of

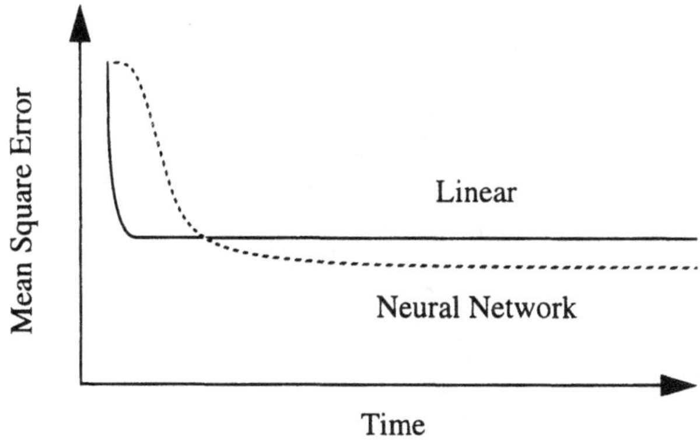

Figure 2: Prototypical result of neural network performance versus a linear equalizer with the same number of taps.

researchers [4][7][9][10][11] and virtually all produce results similar to Figure 2. The key features are that the neural network performs at least as well as the linear equalizer although it takes somewhat longer—in some cases more than an order of magnitude longer—to achieve this better performance. This long acquisition time is less problematic in static applications such as modems or microwave links but may pose a problem for more dynamic environments. Also, in cases plotting symbol error rate as opposed to mean square error, the improvement of the neural network often proved less dramatic.

An interesting result in [9] and produced in simulations by this author is that although for a given number of taps the neural network performs better, a linear equalizer with a greater number of taps would produce better results than both of them. For very large number of taps the neural network often would only produce marginally better results.

4 Recommendations

Part of the problem in many of these comparisons, is that the neural network is trying to compete with a channel that is well understood from the linear channel model point of view, and contains only slight nonlinearities. Problems that contain stronger nonlinearities are described in [1][3]. Such problems would be more convincing evidence of neural networks as enabling technology.

The relatively slow neural network convergence noted in the previous section can be addressed on two fronts. On one, we note that most multilayer perceptrons could easily have direct connections from input to output in addition to the hidden unit connections as in Figure 3. Figure 4 shows results from such a three-layer structure with and without the direct connections as well as from a

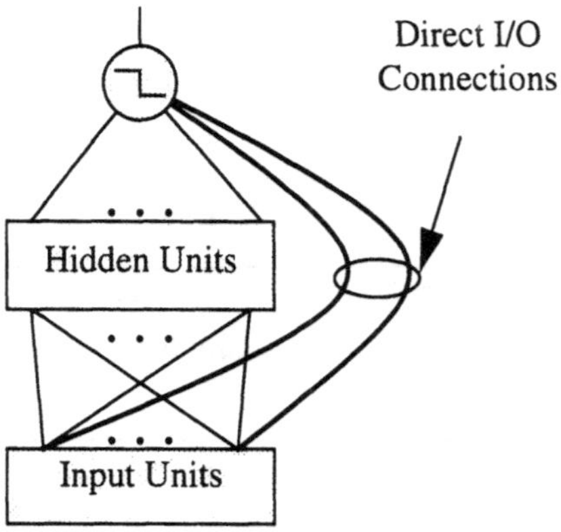

Figure 3: Direct linear connections to speed initial learning.

linear structure (i.e. only the direct connections). These used a simple linear channel model, but still shows that the structure of Figure 3 converges at least as fast as the linear equalizer and faster than the usual three-layer structure. Figure 4 is a similar plot but it shows a channel which can not be compensated for using a feedforward linear equalizer.

On the second front, whereas many of the studies were mainly concerned with error performance, they used standard backprop on-line learning approaches. Of course more sophisticated algorithms exist, and if possible a history of training samples could be maintained so that more powerful batch learning processes be applied. For example, we used the cascade correlation learning

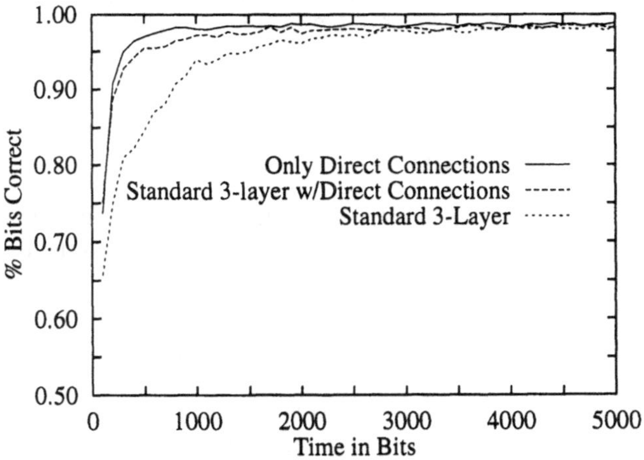

Figure 4: Comparing neural equalizers with different connectivity. Average of 50 runs using 7 taps, delay 3, SNR 20dB, on additive noise linear channel $0.3482 + 0.8704z^{-1} + 0.3482z^{-2}$.

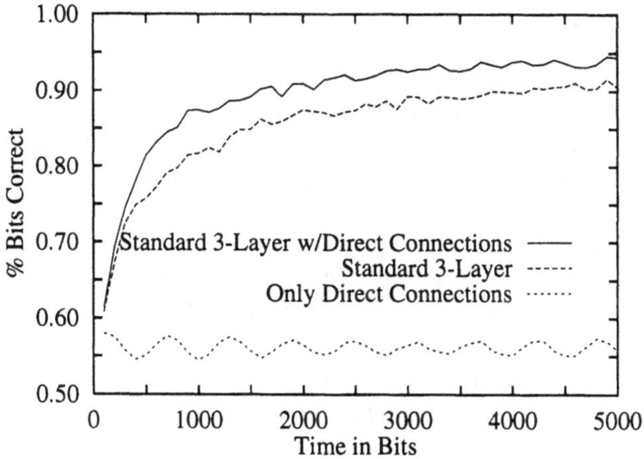

Figure 5: Comparing neural equalizers with different connectivity. Average of 50 runs using 2 taps, delay 0, SNR 20dB, on additive noise linear channel $0.5 + 1.0z^{-1}$.

algorithm [6] to train on all of the sample received so far. Cascade correlation is a batch learning process with two key features: it uses direct connections; and it adds hidden units as needed. Some results of this are shown in Figure 4. This method is computationally intensive with each sample being shown more than 1000 times in some cases, but it clearly shows the improvements that can be gained via more sophisticated learning.

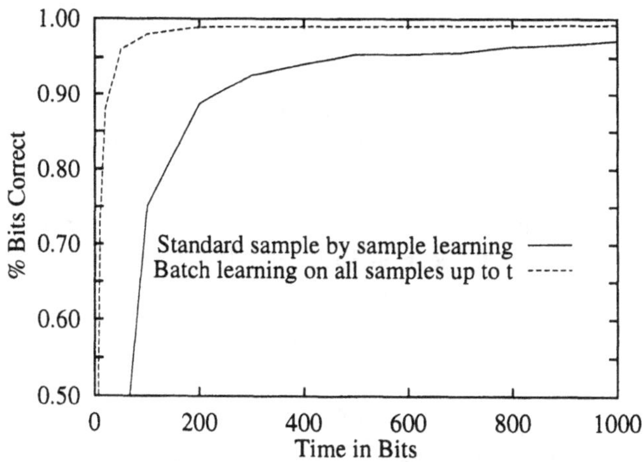

Figure 6: Comparing sample-by-sample learning with batch learning on all of the samples received so far using the channel of Figure 4. Both examples used a network with direct connections.

5 Conclusions:

Neural networks look promising as a method for increasing accuracy over noisy channels suffering inter-symbol interference and other degradations. Analog neural hardware can equalize at rates equivalent to the best digital hardware, but at significantly less power. Simulation results in the literature show that indeed neural networks can give improved performance over linear equalizers but at the expense of increased equalizer acquisition times. This apparent trade-off is not necessary since more sophisticated learning and adding direct connections bring the neural network acquisition time to that of the linear equalizer. Simulation results have verified the method. While arguments exists for using the non-linear neural network, more challenging non-linear channels must be addressed before the enabling characteristics of neural networks can be shown.

References

[1] Agazzi, O., Messerschmitt, D.G., "Nonlinear Echo Cancellation of Data Signals," *IEEE Trans. on Communications*, Vol. COM-30, No. 11, pp. 2421–2433, Nov. 1982.

[2] Alspector, J., Jayakumar, J., Luna, S., "Experimental Evaluation of Learning in a Neural Microsystem," in *Neural Information Processing Systems 4*, ed. Moody, J.E., Hanson, S.J., Lippmann, R.P., Morgan Kauffman, San Mateo CA, 1992, pp. 871–878.

[3] Ariyavisitakul, S., "Equalization of a Hard-Limited Slowly-Fading Multipath Signal Using a Phase Equalizer with Time-Reversal Structure," *IEEE Journal on Selected Areas in Communications*, Vol. 10, No. 3, pp. 589–598, Apr. 1992.

[4] Bang, S.H., Sheu, B.J., "Neural Network Communications Receiver Based on the Nonlinear Filtering," *Proc. IEEE/INNS International Joint Conference on Neural Networks*, Vol. II, pp. 999–1004, 1992.

[5] Chen, S., Gibson, G.J., Cowan, C.F.N., "Adaptive Channel Equalisation Using a Polynomial-Perceptron Structure," *IEE Proceedings*, Vol. 137, Pt. I, No. 5, pp. 257–264, Oct. 1990.

[6] Fahlman, S.E., Lebiere, C., "The Cascade Correlation Learning Architecture," in *Neural Information Processing Systems 2*, ed. Touretzky, D., Morgan Kauffman, San Mateo CA, 1990, pp. 524–532.

[7] Kaneda, R., Manabe, T., Fujii, S., "Adaptive Decision-Feedback Equalizer Using Forward-Only Counterpropagation Networks for Rayleigh Fading Channels," *Proc. of IEEE Neural Networks for Signal Processing II*, pp.

570–578, 1992.

[8] Kirkland, W.R., Taylor, D.P., "On the Application of Feed Forward Neural Networks to Channel Equalization," *Proc. IEEE/INNS International Joint Conference on Neural Networks*, Vol. II, pp. 919–924, 1992.

[9] Kuczynski, W.L., Lao, W., Dong, A.M., Wong. B.C., Nicholas, H.T., Itri, B.A., Samueli, H., "A 1Mb/s Digital Subscriber Line Transceiver Signal Processor," *Proc. IEEE International Solid-State Circuits Conference*, pp. 26–27, 1993.

[10] Lin, J.-Y., Wei, C.-H., "A New Adaptive Equalizer for Nonlinear Channels," *Proc. IEEE/INNS International Joint Conference on Neural Networks*, Vol. II, pp. 981–986, 1992.

[11] Lo, N.W.K., Hafez, H.M., "Neural Network Channel Equalization," *Proc. IEEE International Conference on Communications*, Vol. II, pp. 981–986, 1992.

[12] Lucky, R.W., "Automatic Equalization for Digital Communications," *Bell Syst. Tech. J.*, pp. 547–588, April 1965.

[13] Qureshi, S., "Adaptive Equalization," *IEEE Communications Mag.*, pp. 9–16, Mar. 1992.

Applications of learning techniques to Network Management

Rodney M. Goodman and Barry Ambrose

Electrical Engineering Dept.,
California Institute of Technology,
Pasadena, California 91125 USA
rogo@electra.caltech.edu

Abstract

In this paper we describe the application of learning techniques to some of the network management tasks associated with a system called NOAA (Network Operators Advice and Assistance). NOAA is an expert system that has been successfully developed to aid network management of the Pacific Bell Southern California Telephone Network. Conclusions are drawn regarding the utility and effectiveness of the learning techniques.

1 Introduction

Pacific Bell and CalTech have for some time been working on a real-time traffic management/expert system. This project is called NOAA, Network Operators Advice and Assistance. The task of NOAA is to take information from the Pacific Bell network management computer, use it to isolate and diagnose exceptional events in the network and then recommend the same corrective advice as network management staff would in the same circumstances.

The development of the expert system has been reported in previous papers [1, 2]. Applications of learning techniques to two datasets to facilitate the tasks of network management associated with NOAA are described in this paper.

Dataset	Method	Aim	Comments
Occupancy	linear predictor	time series prediction	neural net better
	neural network	time series prediction	need good choice of number of hidden units
	hidden markov model	detection of non-stationarities	worked well
	hidden markov model	detection of time series increases	worked well
Overflow	linear predictor	time series prediction	no benefit over simple heuristic
	neural network	time series prediction	benefit over simple heuristic
	fourier analysis	detection of periodicity	results not easy to interpret
	count of periodic cases	detection of periodicity	worked well

Table 1: Learning Techniques Applied

Although the problems described in this paper are concerned with monitoring telephony traffic, the techniques should be applicable to the monitoring of any large network, with little modification. For example, rather than monitor trunk overflow, the network manager may be monitoring link error rate, but the same analysis techniques should apply.

The two datasets examined in this paper are:

- Readings of occupancy of a trunk group taken over 5 days.
- Readings of final route overflow on all routes in the network taken over 70 days.

Table 1 summarizes the learning methods applied to the data sets.

2 The Occupancy Dataset

2.1 Introduction

The primary goal of NOAA is to monitor the Southern California telephone network and take appropriate actions if exceptional conditions are diagnosed. In the case of overloads of a single trunk group in the network, the action may be to reroute that traffic elsewhere in the network. To do this, NOAA needs to know where spare capacity exists. This trunk group data is made available to NOAA every 5 minutes in the form of a collection of statistics for each trunk group in the network.

The statistic that is of most use for assessing spare capacity on a trunk group is the occupancy statistic. It is expressed as a percentage of the capacity of a trunk group that is taken up by traffic. In this paper, a case study is carried out of the application of neural networks to time series prediction of the occupancy statistic. Linear predictors are used as a reference for comparison. Since neural networks can learn arbitrarily complex functions, it is to be expected that neural networks will do better that linear predictors in learning to predict occupancy.

We used a training file of 1200 examples and a test file of 300 examples. The data represented occupancy on a trunk group and we were trying to predict future occupancy based on the previous six observations. Using the neural network, we trained until the errors on the test set rose 0.1% above their minimum level.

We decided to use the Quickprop [4] program for network training as it was advertised as having faster convergence than standard backprop. The quickprop program incorporates a weight decay factor which avoids overtraining. We modified it to include linear outputs since squashing functions on the output units will not aid function fitting.

A range of 3 to 10 hidden units were used. We tried to quantify the benefit of using a neural network by examining performance on the test set and we recorded the minimum test set error. The data was shuffled 6 times to generate 6 runs, to get a better measure of the variation from case to case.

The results are shown in Table 2 averaged over the six runs. Note that the neural network always showed a significant advantage over the linear predictor on the training set, so Table 2 just lists performance on the test set. The results show that the neural network with four hidden units did significantly better than the linear predictor assuming a 90% statistical confidence level. The performance comparison with more or less hidden units showed no statistically significant advantage or disadvantage.

Table 3 shows the variation in results on a single training set when using a different random seed for the neural network to start with. Five runs on the same data were used. This illustrates an extra source of uncertainty in the overall comparison since we may find a local rather than global minimum (as evidenced by Table 3).

The average occupancy reading was about 0.2. The RMS error for the linear predictor was 0.015.

The results indicate that neural networks can learn the time series pre-

Hidden Units	RMS Benefit	Standard Deviation
3	−0.000267	0.000401
4	0.000204	0.000261
6	0.000120	0.000241
10	−0.000011	0.000146

Table 2: Decrease in RMS error for Neural Net compared to Linear Predictor

Hidden Units	Standard Deviation
3	0.000272
4	0.000384
6	0.000620
10	0.000315

Table 3: Uncertainty due to local vs. global NN minimum

diction task slightly better than linear predictors.[1] However there is a danger of overfitting, that is, learning the noise as well as the underlying function if too many hidden units are present. Also training of linear predictors is a lot less computationally intensive than training of neural networks. This may give an advantage to linear predictors in situations where predictions have to be carried out in real time.

2.2 Hidden Markov Models

Following on from this work, we experimented with hidden Markov Models. What follows is based on techniques and theory developed by Dr. P. Smyth of JPL.

Suppose that we are examining a time series $x_1, x_2, x_3 \ldots$ and based on the latest reading x_n we wish to classify the underlying process as being in one of three states. The first approach we could take is to make a table of all readings and find the probability distribution of each of the states. In other words, from past experience, given a reading x_n, we could associate a probability $p(s_i|x_n)$ with that reading, the probability of the random process being in state i given a reading of x_n.

A better approach that can often be justified is to use information from the previous readings $x_{n-1}, x_{n-2}, \ldots$ to increase the confidence of the classification. In particular, if we assume the underlying process is Markov in

[1] Note that making the usual assumptions for telephone traffic, i.e. memoryless arrivals and holding times, and assuming stationarity, it can be shown theoretically that a linear predictor will be close to optimal for prediction of occupancy, since the occupancy process can be approximated as a stationary joint Gaussian stochastic process [3].

nature, we can associate a transition matrix A with each time step. The transition matrix is made up of entries a_{ij}, where a_{ij} represents the probability of transition to state i from state j.

Based on this approach, a better classification technique is to set

$$p_1^n = p(1|x_n) \times (a_{11}p_1^{n-1} + a_{12}p_2^{n-1} + a_{13}p_3^{n-1})$$
$$p_2^n = p(2|x_n) \times (a_{21}p_1^{n-1} + a_{22}p_2^{n-1} + a_{23}p_3^{n-1})$$
$$p_3^n = p(3|x_n) \times (a_{31}p_1^{n-1} + a_{32}p_2^{n-1} + a_{33}p_3^{n-1})$$

where p_i^n is our estimate of the relative probability of being in state i at timestep n.

We tried two Markov models. The first looked at the time series of prediction errors from the linear predictor and tried to detect non-stationarities. Thus we presumed the trunk occupancy process was in one of two states, either stationary where the linear predictor was doing a good job, or non-stationary, where the linear predictor gave large errors. Given that telephone traffic is generated for the most part by humans who may share a large interest in particular events, (e.g. Monday night football), it is to be expected that trunk occupancy should exhibit occasional non stationarities when these events occur.

The statistics of the prediction error showed a normal distribution except for a large concentration about zero error and the presence of some outliers at a distance of 4 or more standard deviations from zero. The zero error cases can be explained by night time traffic where both the traffic and the error were near zero for 8 hours at a time, and the outliers can be explained by the non-stationarities that we are trying to detect.

Based on some experimentation, the stationary distribution was set to be a normal curve peaking at $p(s = 1|x_n = 0) = 0.9$ and with a standard deviation equal to the prediction error standard deviation. The non stationary distribution was set to be $p(s = 2|x_n) = 1 - p(s = 1|x_n)$. Examples of the classification are given in Figure 1 and Figure 2. A transition probability of 0.99 for staying in the same state at the next time step was used for each state.

Classification of the occupancy process into stationary or non-stationary categories could be useful during training of the linear predictor. If a non-stationary set of examples is encountered, training could be turned off or a different linear predictor trained. The neural network on the other hand could be expected to automatically use its hidden units to capture the dynamics of the non-stationary mode.

The second use of hidden markov models for classification looked at the jump statistic $x_n - x_{n-1}$ to try to detect increases or decreases in the average traffic level. An increase in the level could be indicated by a series of large

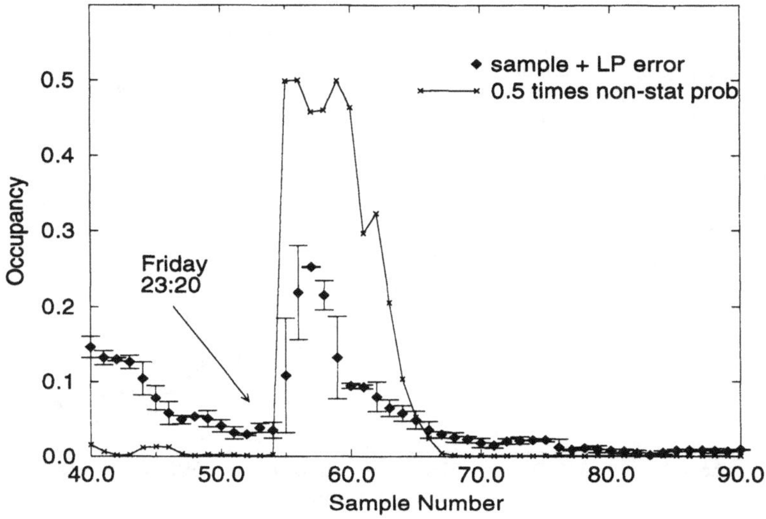

Figure 1: Detection of non-stationarity using a Markov Model

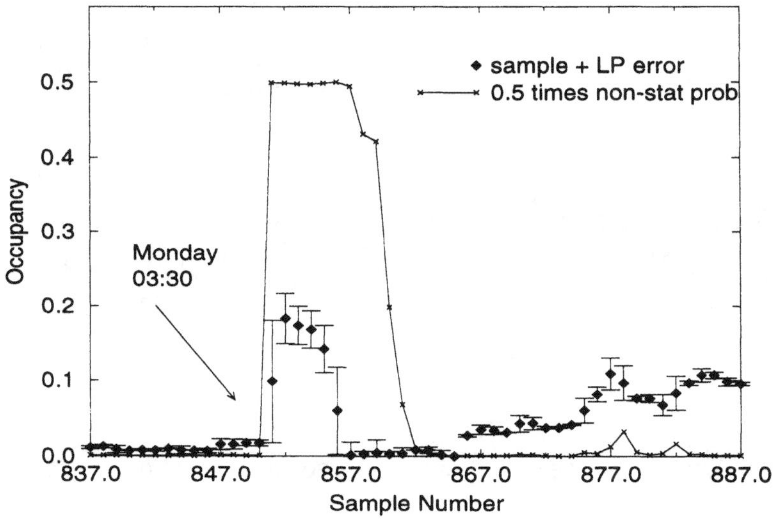

Figure 2: Detection of non-stationarity using a Markov Model

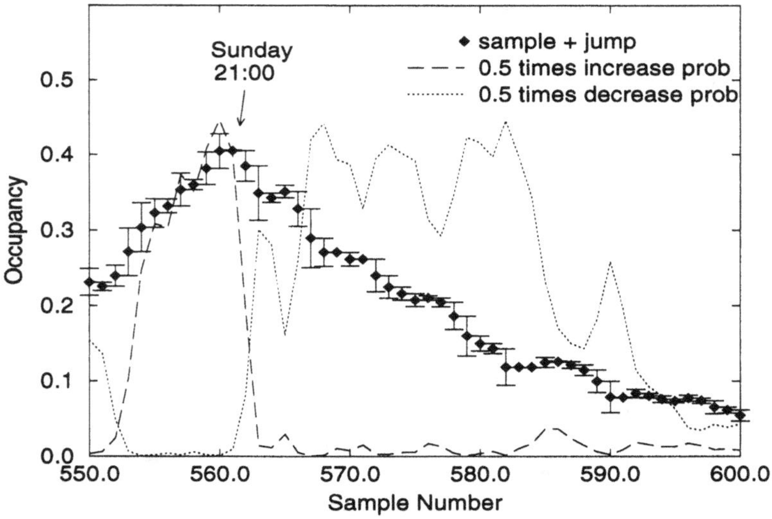

Figure 3: Detection of increase/decrease using a Markov Model

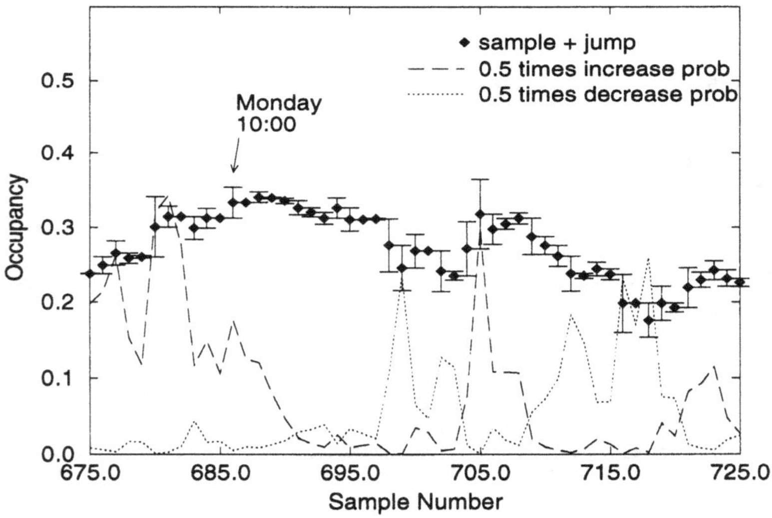

Figure 4: Detection of increase/decrease using a Markov Model

positive and small negative jumps for example. This classification could be useful in deciding whether to use a trunk route for rerouting traffic. If the trunk group is in an increasing state, its available capacity could be derated by some factor. Conversely if it is in a decreasing state, its capacity could be multiplied up by the same factor. If the trunk level is steady state, we can use the observed capacity without modification.

The statistics of the jumps showed a normal distribution with a concentration about zero and some outliers for the same reasons as given for the prediction error distribution.

Based on some experimentation, the "no change" distribution was set to be a normal curve peaking at $p(s = 1|x_n = 0) = 1.0$ and with a standard deviation equal to the jump standard deviation. The "increasing" distribution was set to be a normal curve peaking at one standard deviation above zero. The choice of one standard deviation was a compromise. More than one standard deviation and only very large increases would be spotted. Less than one standard deviation and there would be too much confusion with the "no change" classification. The probability of the "decreasing" distribution was set to be a normal curve peaking at one standard deviation below zero. The probabilities were then normalized so that they summed to one.

Examples of the classification are given in Figure 3 and Figure 4. A transition probability of 0.96 for staying in the same state at the next time step was used for each state.

3 The Overflow Dataset

3.1 Introduction

In the second case study of the utility of neural networks in the network management environment, we examined network wide statistics of overflow. Rather than dealing with each exception as an isolated event, which NOAA currently does except in very special circumstances, events that are correlated in space and time can be detected and more appropriate actions taken. We expected to see some patterns in the large volume of network management data, which the managers were not aware of, and which can be used to aid decision making.

When putting controls into the network, it is useful to predict whether the overflow is once off or whether it is likely to reoccur. Here we associate a cost with putting a control into the network. The cost is 1 call unit for each 5 minute period the control is in the network. We wish to predict the expected value of overflow in the next five minutes period, given its value in the last five minutes and put a control into the network if the expected value of overflow is greater than the control cost.

3.2 Linear Predictor

We first tried using a linear predictor on the data. We considered 300 routes that had between 20 and 200 overflow occurrences on them over the 70 day period. The data represented calls overflowing and we were trying to predict future overflow based on the previous six readings. After looking at some sample data, we found that the best way of training for the all zero data case which made up the majority of function examples was to take out the constant term in the linear predictor network.

A heuristic rule, similar to that currently implemented in NOAA was used as a basis for comparison. This rule gives good performance in guiding us as to when to put in or remove a control (which has a cost of one call unit when implemented). The rule was:

> If the current overflow reading is zero, expected value of next overflow reading is zero. If the current overflow reading is one or more, the expected next reading of overflow is one or more.

We divided the data for each route into a test set and a training set and then checked for benefit on the test set only. We stopped training when the error on the test set rose 0.1% above its minimum value. This avoided over-training.

We found the average benefit of the linear predictor over the heuristic rule to be 14.74 with a standard deviation of 203.9. We can check whether this is statistically different from 0 at a 95% confidence level:

$$\begin{aligned} z &= \frac{14.74}{\sqrt{41575/300}} \\ &= 1.25 \end{aligned}$$

which is not statistically significant, i.e. there is no clear statistical advantage of learning compared to the simple heuristic rule on the 300 test cases.

The linear predictor may not have done well for two reasons:

- There are not a lot of examples from which to learn the function, however the linear predictor should come up with a function that is in some sense conservative, i.e. its function space is not rich.
- The function is not expected to be a linear mapping. This is the most likely reason that we did not see a lot of benefit. The overflow time series saturates on the lower end of the scale at zero. Thus its estimator is likely to be non linear.

3.3 Neural Network

We now try a neural network on the same task. The classification we are trying to learn is a Boolean one, that should be set true if the expected value of overflow given previous inputs is > 1. The all zero's test case needs proper handling. This is done by modifying the program to present the all zero's case after each non-zero example. Again training is carried out until the error on the test set begins to rise.

We associated a target of -20 with the all zero's case, and a target of the overflow level with the non-zero data, although a 0/1 squashing output unit was used. This was to associate a large error penalty with the case of large overflow. A fixed number of iterations was used, since the training set error showed only a slow decrease.

We found an average benefit is 20.0467 with a standard deviation of 159.7. We can check whether this is statistically different from 0 at a 95% confidence level:

$$
\begin{aligned}
z &= \frac{20.0467}{\sqrt{25514.6/300}} \\
&= 2.17
\end{aligned}
$$

which is statistically significant, i.e. there is a clear statistical advantage of the neural network compared to the simple heuristic rule on the 300 test cases.

Looking at some examples of the weights of the neural networks, an "OR" function across all the inputs seemed to be implemented.

3.4 Detecting Periodicity

An obvious aid to prediction of overflow is detecting periodicity in the data. We took 300 routes that had between 20 and 200 overflow occurrences over the seventy day test period and checked for periodicity either on a 24 hour basis or a weekly basis. Two methods were used. A Discrete Fourier Transform (DFT) was carried out to look for harmonics that signified a 24 hour or weekly period and secondly a voting or counting approach was used. For each data sample, 30 votes were accumulated if there was an overflow presently and an overflow condition 24 hours later. One vote was accumulated if there was no overflow presently and no overflow 24 hours later. The routes with the highest score were picked for visual examination.

The DFT did not prove useful as the overflow "bursts" contain energy at a range of low frequencies, including the harmonics that we were interested in. Visual inspection of the signals with the highest energy at the harmonics we were interested in showed no obvious periodicity. On the other hand, the routes picked by the voting method showed clear periodicity, which could be used for predictive purposes.

The voting or counting procedure could probably be used also to detect correlations in space (e.g. one route always overflowing when another one does). A matrix of routes would be needed to hold the votes.

4 Conclusions

In this paper, a comparison of traditional prediction techniques with a neural network approach has been given. Neural networks using back propagation have again shown their utility in function estimation and learning from data examples. However the correct number of hidden units have to be used for best results.

5 Acknowledgements

This work is supported in part by Pacific Bell, and in part by DARPA and ONR under grant no. N00014-92-J-1860.

References

[1] Goodman, R. M., Ambrose, B., Latin, H., Finnell, S., "Network Operators Advice and Assistance (NOAA): A real-time traffic rerouting expert system," *Globecom,* Florida, December, 1992.

[2] Goodman, R. M., Ambrose, B., Latin, H., Finnell, S., "Network Operators Advice and Assistance (NOAA): A hybrid Neural Network / Expert System for Traffic Management," *IFIP,* San Francisco, April, 1993.

[3] Christensen, R., *Linear Models for Multivariate, Time Series, and Spatial Data,* Springer-Verlag, New York, 1991.

[4] Fahlman, S. E., "Faster-Learning Variations on Back-Propagation: An Empirical Study" in *Proceedings of the 1988 Connectionist Models Summer School,* Morgan Kaufman, 1988.

A Distributed Reinforcement Learning Scheme for Network Routing

Michael Littman*, Justin Boyan

Carnegie Mellon University
School of Computer Science
Pittsburgh, PA
** also Cognitive Science Research Group, Bellcore*
Morristown, NJ

Abstract

In this paper we describe a self-adjusting algorithm for packet routing in which a reinforcement learning method is embedded into each node of a network. Only local information is used at each node to keep accurate statistics on which routing policies lead to minimal routing times. In simple experiments involving a 36-node irregularly-connected network, this learning approach proves superior to routing based on precomputed shortest paths.

1 Introduction

An algorithm which can monitor its own performance has the possibility of adapting its behavior to situations unforeseen by its human designers. In this paper, we present an example of such an algorithm for the problem of routing packets efficiently in a communication network with an irregular topology and unpredictable usage patterns [4, 5]. The algorithm must learn a routing policy which balances minimizing the number of "hops" a packet will take with the possibility of congestion along popular routes. It does this by experimenting with different routing policies and collecting statistics on which policies minimize total delivery time.

Although in principle such an approach could be very expensive in terms of learning time and storage space, the algorithm we describe here is a variant of a reinforcement learning algorithm called *Q-learning* [7] which adapts to changes in network traffic and requires little more space than that needed to represent a complete routing policy. The application of Q-learning here differs from its traditional use in that the network is actually constructed from a distributed *collection* of learners, each of which is responsible for a portion of the problem. This approach appears to be very effective in routing packets efficiently under high load.

The experiments in this paper were carried out using a discrete event simulator to model the motion of packets through a local area network. Packets are periodically introduced into the network at a random node with a random destination. Multiple packets at a node are stored in an unbounded FIFO queue; however, we set a limit on the total number of packets active in the network at a time, generally 1000. In unit time, a node takes the top packet in its queue, examines its destination, and chooses a neighboring node to which to send the packet. A packet sent directly to its destination node is removed from the network immediately.

2 Routing as a Reinforcement Learning Task

A packet routing policy answers the question: to which adjacent node should the current node send its packet in order to get it as quickly as possible to its eventual destination? Since the policy's performance is measured by the total time taken to deliver a packet, there is no "training signal" for directly evaluating or improving the policy until a packet finally reaches its destination. However, using an idea from the field of reinforcement learning, we can update the policy more quickly and using only local information.

In our learning scheme, the policy is distributed throughout the network as follows: each node keeps an estimate, for every neighbor/destination pair (y, d), of how long it should take for a packet with destination d to arrive there if it is sent first to neighbor node y. When a node x is asked to route a packet, it sends it to that neighbor $\bar{y}$ which x estimates will have the lowest total delivery time[1]. Instead of then waiting for the packet to finally reach d before updating the policy, x simply queries $\bar{y}$ to find out how long *it* expects the given packet to take in getting to d. Since the neighboring node is presumably closer to the final destination, its estimate is considered

[1] In our experiments, we had nodes occassionally send a packet to a random neighbor to help the system locate shortcuts.

more accurate and thus can be used to update the sending node's delivery time estimate.

More precisely, let $Q_x(\bar{y}, d)$ be the time that node x estimates it takes to deliver a packet P bound for node d by way of x's neighbor node $\bar{y}$, including any time that P would have to spend in node x's queue.[2] Upon sending P to $\bar{y}$, x immediately gets back $\bar{y}$'s estimate for the time remaining in the trip, namely

$$Q_{\bar{y}}(\bar{z}, d) = \min_{z \in \text{neighbors of } \bar{y}} Q_{\bar{y}}(z, d)$$

If the packet spent q units of time in x's queue, then x can revise its estimate as follows:

$$\Delta Q_x(\bar{y}, d) = \eta \, (\overbrace{Q_{\bar{y}}(\bar{z}, d) + q}^{\text{new estimate}} - \overbrace{Q_x(\bar{y}, d)}^{\text{old estimate}} \,)$$

where η is a "learning rate" parameter (0.7 in our experiments).

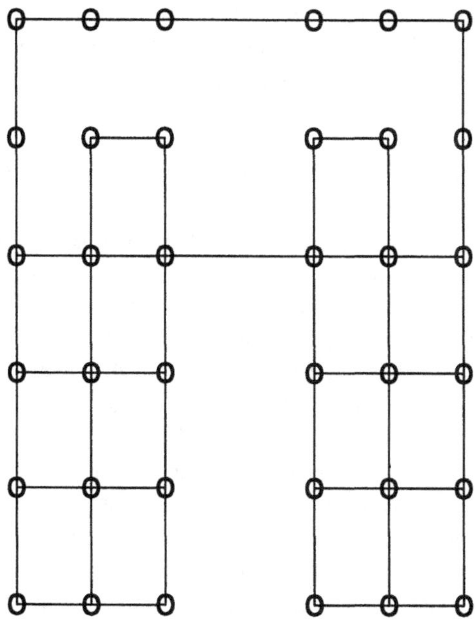

Figure 1: Network topology

In the field of reinforcement learning, the Q-function, $Q_x(y, d)$, is often approximated by a neural network (see e.g. [3, 6]); this can allow the learner

[2]We denote the function by Q because it corresponds to the "Q-function" used in the reinforcement learning technique of Q-learning [7].

to incorporate diverse parameters of the system, such as local queue size and time of day, into its distance estimation. For the implementation described here, however, we represented Q as a table.

3 Results

We tested our routing algorithm on a variety of network topologies, including the 7-hypercube, a 116-node LATA telephone network, and an irregular 6×6 grid. Varying the level of network traffic, we measured the average delivery time for packets in the system after learning had settled on a routing policy, and compared these delivery times with those given by a conventional routing scheme based on shortest paths. The result was that in all cases, the learning algorithm was able to sustain a higher level of network traffic than the non-learning one.

Here we present the results for the irregular grid network (pictured in Figure 1). Under conditions of low load, the network learns fairly quickly to route packets along shortest paths to their destination.[3] The performance vs. time curves plotted in Figure 2 demonstrate that our routing algorithm, using no prior knowledge of the network topology, learns to route about as well as the shortest path router, which performs optimally in low load (left picture).

As network traffic increases, however, the shortest path routing scheme is far from optimal: it ignores bottlenecks and soon floods the network with packets. The right half of Figure 2 plots performance vs. time for the two routing schemes under high load conditions: while shortest path is unable to tolerate the packet load, our algorithm learns an efficient routing policy. The reason for the learning algorithm's success is apparent in the "policy summary diagrams" in Figure 3. These diagrams indicate, for each node under a given policy, how many of the 36×35 point-to-point routes go through that node. On the left side of Figure 3, which summarizes the shortest path routing policy, two nodes in the center of the network (labelled 570 and 573) are on many shortest paths and thus become congested when network load is high. By contrast, the diagram on the right shows that our algorithm, under conditions of high load, learned a policy which routes some traffic over a longer than necessary path (across the top of the network) so as to avoid congesting the nodes in the center of the network.

Our basic result is captured in Figure 4, which compares the performances of the shortest path policy and learned policy as the network load increases. Each point represents the median over three trials of the mean

[3] In fact, it is interesting to note that the Q-learning update rule is mathematically very much like the well-known Bellman-Ford shortest paths algorithm [1, 2], except our path relaxation steps are performed asynchronously.

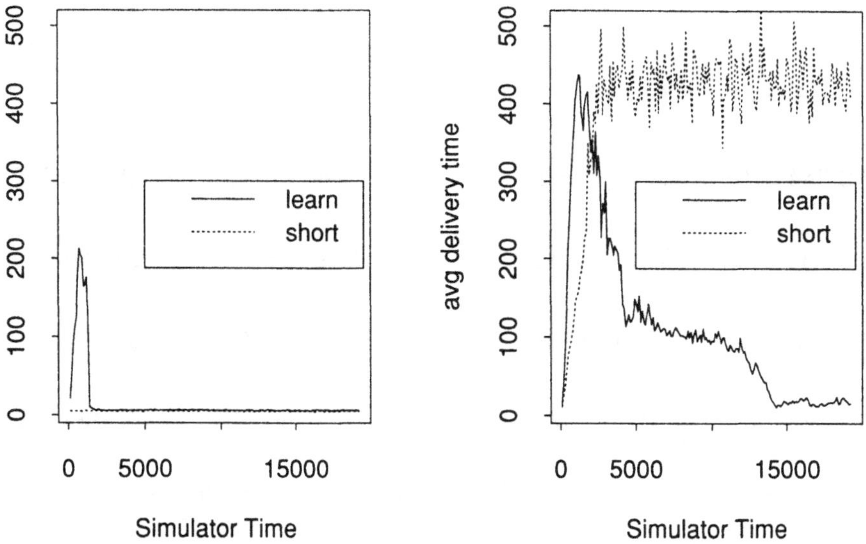

Figure 2: Performance under low load and high load

packet delivery time (after learning has settled). When the load is very low, our learning algorithm routes nearly as efficiently as the shortest path policy. As load increases, the shortest path policy leads to exploding levels of network congestion, whereas the learned policy continues to route efficiently. Only after a further significant increase in load does the distributed learning algorithm, too, result in congestion.

4 Conclusion

In this paper, we have exhibited a learning algorithm which, without having to know in advance the network topology and traffic patterns, and without the need for any centralized routing control system, is able to discover an

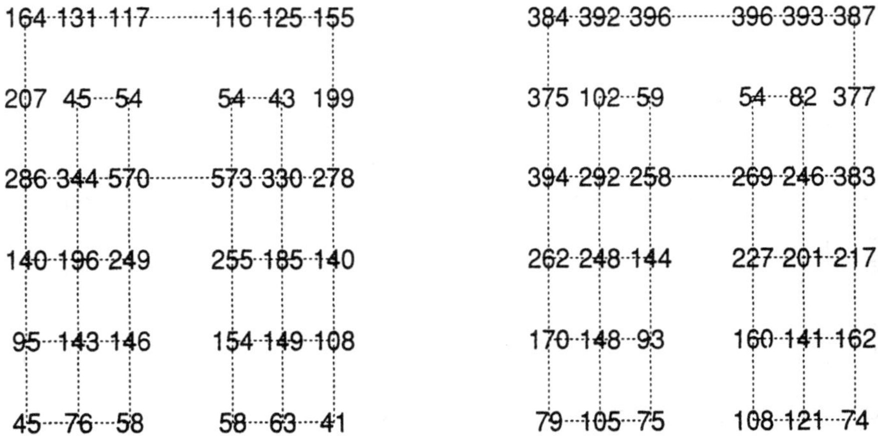

Figure 3: Policy summaries: shortest path and learned under high load

efficient routing policy. Although the simulations described here are not fully realistic from the standpoint of actual telecommunication networks, we believe that reinforcement learning algorithms such as this one could be effective for self-adapting routing and other practical tasks.

References

[1] R. Bellman. On a routing problem. *Quarterly of Applied Mathematics*, 16(1):87–90, 1958.

[2] L. R. Ford, Jr. *Flows in Networks.* Princeton University Press, 1962.

[3] L.-J. Lin. *Reinforcement Learning for Robots Using Neural Networks.* PhD thesis, School of Computer Science, Carnegie Mellon University, 1993.

[4] H. Rudin. On routing and delta routing: A taxonomy and performance comparison of techniques for packet-switched networks. *IEEE Transactions on Communications*, COM-24(1):43–59, January 1976.

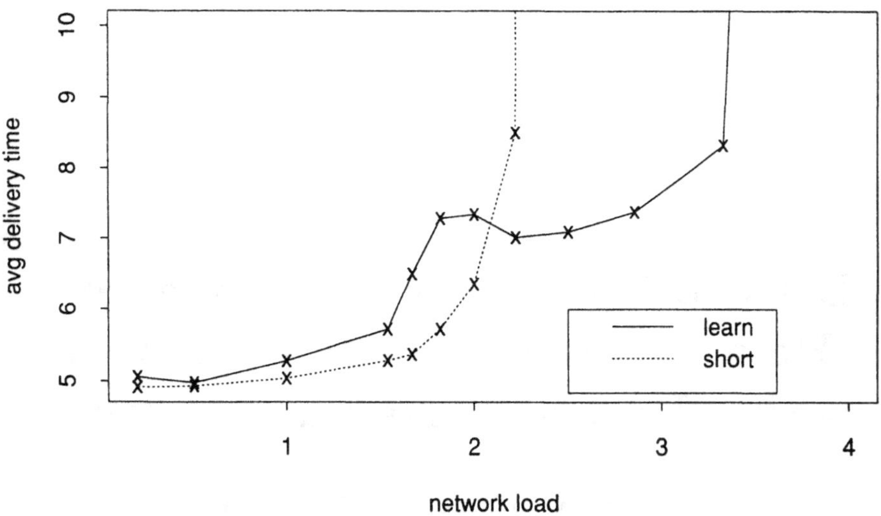

Figure 4: Delivery time at various loads

[5] A. Tanenbaum. *Computer Networks.* Prentice-Hall, second edition edition, 1989.

[6] G. Tesauro. Practial issues in temporal difference learning. *Machine Learning*, 8(3/4), May 1992.

[7] C. Watkins. *Learning from Delayed Rewards.* PhD thesis, King's College, Cambridge, 1989.

Discovering the Structure of a Self-Routing Interconnection Network with a Recurrent Neural Network [1]

Mark W. Goudreau* and C. Lee Giles

NEC Research Institute
4 Independence Way, Princeton, NJ 08540
giles@research.nj.nec.com
**currently at Dept. of Computer Science*
U. of Central Florida, Orlando, FL 32816
goudreau@cs.ucf.edu

Abstract

A Recurrent Neural Network (RNN) is used to *learn* the structure of an Interconnection Network (IN) by training with examples of routing messages. This is essentially the same as using RNNs to learn the structure of a synchronous sequential machine for which there are many distinct initial states. The method is illustrated by learning a small Self-Routing Interconnection Network (SRIN) with 6 switches by training a second order Single Layer RNN (SLRNN).

1 Introduction

Interconnection Networks (INs) are usually *designed* not learned. (However, one could say that the IN designer at some point did have to learn how to design INs.) Engineers have designed INs based on certain logical structures for INs and their eventual use. Many different criteria can be considered, i.e. speed, complexity, and ease of route calculation [12].

[1] This paper is adapted from Chapter 6 of reference [3].

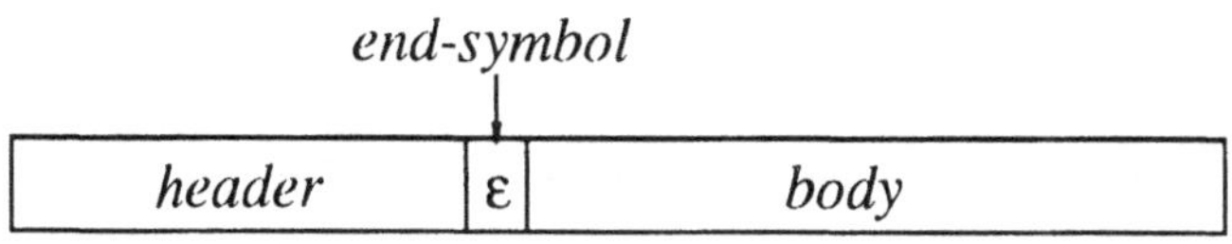

Figure 1: A SRIN message where the header and body are strings of symbols. Separating the header from the body is an end-symbol, ϵ.

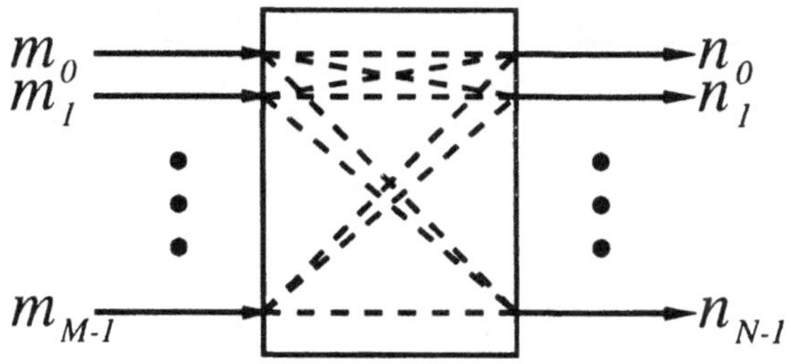

Figure 2: An $M \times N$ self-routing switch where input messages are routed on a First-In, First-Out (FIFO) basis.

Though there has been much work on using neural networks for routing interconnection networks (see for example [1, 4, 8, 9, 11, 13]), nothing, as far as we know, has been done on learning INs with neural networks. (A related paper on the learning of INs using simulated evolution is Hillis [6].) We discuss the learning of Self-Routing Interconnection Networks (SRINs) using Recurrent Neural Networks. SRINs can be used to describe many commonly used INs. Consider a parallel computing system that uses message passing to communicate: the processors have certain communication requirements with other processors and certain message headers must be used. We use a second order Single Layer Recurrent Neural Network (SLRNN) to learn the message routing training data. After training, the SRIN is extracted from the SLRNN. Thus, learning a Synchronous Sequential Machines (SSMs) from examples can be seen as quite similar to this process.

2 Self-Routing Interconnection Networks

SRINs permit a set of processors to communicate amongst themselves using a *store-and-forward* methodology. The switches in a SRIN examine the message that is being sent and decide which way to route it. We define a message as consisting of two parts: a *header* and a *body*. The header and the body are separated by an *end-symbol*, ϵ. Figure 1 gives a schematic of such

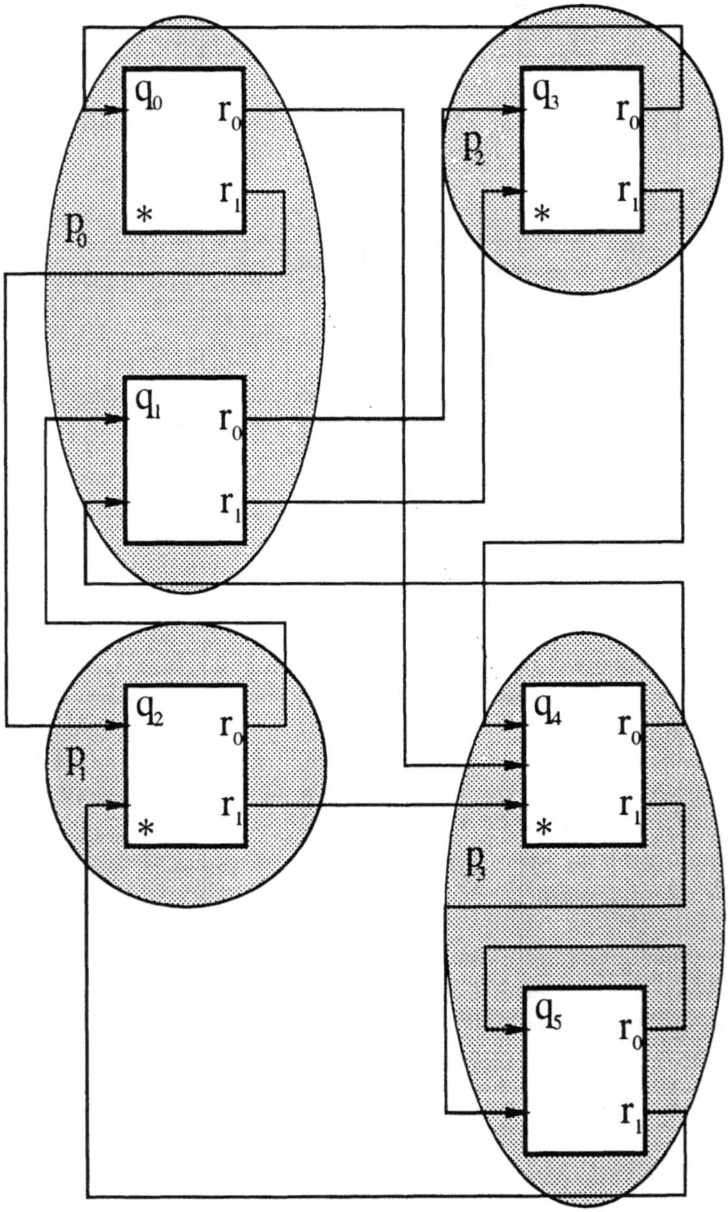

Figure 3: A sample self-routing interconnection network.

a message. *Self-routing switches* is used by the SRIN to route messages. Figure 2 is a drawing of a $M \times N$ self-routing switch.

We give an intuitive explanation of the routing. Suppose a message arrives at a switch. The switch strips off the first symbol of the message header and examines it. If the symbol is not the end-symbol, ϵ, the switch sends the message (minus the header symbol that it just examined) through to the appropriate output port. For the end-symbol, the message is given to the processor that is associated with the switch. Once the header has been stripped down to only the end-symbol, the message is no longer passed through the SRIN. Although it is not shown in Figure 2, there is a connection from each switch to its associated processor, i.e. another output port for the end-symbol, ϵ. The switches work in a First-In, First-Out (FIFO) manner. Messages that can not be routed immediately are buffered until they can be routed.

In Figure 3 we see an example of a SRIN. This SRIN has the set of switches $Q = \{q_0, q_1, q_2, q_3, q_4, q_5\}$. In Figure 3, the switches are shown as white boxes with their labels in the upper left corner. The outputs, labeled r_0 and r_1, are on the right side of each switch. Inputs are at the left side of the switch. Each switch with a * in its lower left corner is a designated switch that starts the message passing procedure for its associated processor. The set of processors is $P = \{p_0, p_1, p_2, p_3\}$. In Figure 3, the processors are represented by shaded areas.

3 Learning Interconnection Networks

To do the training we have for examples list of source processor, header, and destination processor combinations that must be implemented by a SRIN. The task is to infer a SRIN that can accomplish all of the routings described in the training list. In addition, we would like the SRIN to perform correct routings even for examples that are not on the training list. Table 1 contains data for a sample problem and is consistent with the SRIN in Figure 3.

3.1 Single Layer Recurrent Neural Network

We use a second order Recurrent Neural Network for training, see Figure 4. Previous work has shown the advantages of using second order networks for learning finite state automata [5, 10, 14]. Second order means that Neuron j and input k have a combined effect on neuron i modified by the second-order weight w_{ijk}. The output of neuron i is defined as:

$$y_i^t = g(\sum_{j=1}^{N} \sum_{k=1}^{M} w_{ijk} y_j^{t-1} x_k^t) \qquad (1)$$

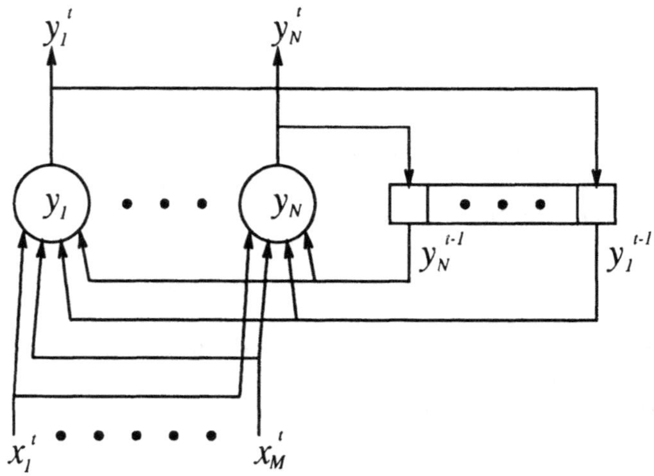

Figure 4: A Canonical form for a Single Layer Recurrent Neural Network (SLRNN) with M input bits, N state bits and (up to) N output bits. A bank of N latches is shown on the right.

The activation function, $g(x)$, is the sigmoid function $g(x) = \frac{1}{1+e^{-ax}}$.

The SLRNN will learn the training data, and the SRIN will be extracted from the SLRNN. A RTRL on-line gradient descent algorithm that is similar to Williams and Zipser [15] was used to train the SLRNN.

The problem of inferring a SRIN, as opposed to inferring a SSM, is that in a SRIN there will generally be several possible initial states, corresponding to the designated switches of different source processors. Thus, the SLRNN that is trying to learn a SRIN will, in fact, be trying to learn a SSM for each possible initial state. However, all of these SSMs should have the same structure; the only difference is that the initial state varies. This is what happens: rather than learning several different SSMs in different sections of the state space, the SLRNN will take advantage of the identical structures of the SSMs and merge them.

3.2 Representation and Training

The input vectors of the SLRNN represent the symbols in the headers of messages that must be routed through the SRIN. The state vectors of the SLRNN represent the switches of the SRIN. And the output vectors of the SLRNN represent the processors of the SRIN.

To train the SLRNN. let the input to the SLRNN be an initial state vector and a series of input vectors applied to the SLRNN at time step one,

source processor	header	destination processor
⋮	⋮	⋮
p_0	r_1	p_1
p_2	$r_1r_1r_0r_0r_1$	p_1
p_2	$r_0r_0r_0$	p_0
p_3	r_0r_1	p_2
p_1	$r_1r_1r_0$	p_3
⋮	⋮	⋮

Table 1: Sample training data for a SRIN which is consistent with the SRIN in Figure 3.

the second input vector at time step two, and so on. The last input vector always corresponds to the end-symbol. The outputs of the neurons are not examined until the entire input has been applied. After the end-symbol is entered, the forward- propagation RTRL is used to train the SLRNN so that the actual output gets closer to the desired output. The state vectors that are created are examined *after* training since these states actually represent switches in the SRIN. Because the neurons have continuous activation functions, we use partitioning and clustering techniques to find equivalent state clusters [2]. [See [16] for a method that seems to avoid these partitioning and clustering methods.] From these clusters a SRIN is extracted from the SLRNN [2] which may or may not be minimal. Standard SSM minimization techniques [7] can be used.

4 Training Simulation

The RNN was trained with training data consistent with the SRIN in Figure 3. The data table contains an entry for every header up to length 11 (including the end-symbol) for each processor and starting with strings of length one and ending with strings of length 11. Table 2 shows how the SLRNN represents the SRIN after training has occurred. This table is actually a condensation of the SRIN switch configuration for 4 different initial switch conditions.

5 Conclusions

We use training data from an unknown SRIN to teach a RNN to eventually represent an unknown SRIN. The trained RNN is capable of routing the

switch	QR
q_0	10000000
	10000001
	10000101
q_1	11000011
q_2	00100000
	00100001
	00100101
	01100101
q_3	00010000
	00010101

switch	QR
q_4	00001000
	01010100
	01011100
	01011110
	01111100
q_5	11000101
	11001101
	11010100
	11011101

Table 2: An extracted representation from the trained SLRNN of the switches of the SRIN shown in Figure 3. QR is the quantized representation of the associated switch in the state space of the SLRNN. The SLRNN has eight neurons.

training data and acting like a SRIN. In addition, a SRIN was *extracted* from the trained RNN that was consistent with the actual SRIN used to generate the training data. As far as we know, this is the first time a neural network has been used to learn a SRIN. It remains to be seen how useful such learning approaches are for real problems.

References

[1] T. Brown and K.-H. Liu, "Neural network design of a Banyan network controller," *IEEE Journal on Selected Areas of Communication*, vol. 8, no. 8, pp. 1428–1438, October 1990.

[2] C. Giles, C. Miller, D. Chen, H. Chen, G. Sun, and Y. Lee, "Learning and extracting finite state automata with second-order recurrent neural networks," *Neural Computation*, vol. 4, no. 3, pp. 393–405, 1992.

[3] M. Goudreau, *Neural Network Applications for Interconnection Networks.* PhD thesis, Princeton University, Princeton, NJ, June 1993.

[4] M. Goudreau and C. Giles, "Routing in random multistage interconnection networks: Comparing exhaustive search, greedy and neural network approaches," *International Journal of Neural Systems*, vol. 3, no. 2, pp. 125–142, 1992.

[5] M. Goudreau, C. Giles, S. Chakradhar, and D. Chen, "First-order vs. second-order single layer recurrent neural networks," *IEEE Transactions on Neural Networks*, 1993. To be published.

[6] W. Hillis, "Co-evolving parasites improve simulated evolution as an optimization procedure," *Physics D*, vol. 42, pp. 228–234, 1990.

[7] J. Hopcroft and J. Ullman, *Introduction to Automata Theory, Languages, and Computation*. Reading, MA: Addison-Wesley Publishing Company, Inc., 1979.

[8] A. Marrakchi and T. Troudet, "A neural net arbitrator for large crossbar packet-switches," *IEEE Transactions on Circuits and Systems*, vol. 36, no. 7, pp. 1039–1041, July 1989.

[9] P. Melsa, J. Kenney, and C. Rohrs, "A neural network solution for call routing with preferential call placement," in *Proceedings of the 1990 Global Telecommunications Conference*, pp. 1377–1382, December 1990.

[10] C. Miller and C. Giles, "Experimental comparison of the effect of order in recurrent neural networks," *International Journal of Pattern Recognition and Artifical Intelligence*, 1993. To be published.

[11] M. Mittler and P. Tran-Gia, "Performance of a neural net scheduler used in packet switching interconnection networks," in *1993 IEEE International Conference on Neural Networks*, vol. II, pp. 695–700, 1993.

[12] H. Siegel, *Interconnection Networks for Large Scale Parallel Processing*. New York: McGraw-Hill, 1990.

[13] T. Troudet and S. Walters, "Neural network architecture for crossbar switch control," *IEEE Transactions on Circuits and Systems*, vol. 38, no. 1, pp. 42–56, January 1991.

[14] R. Watrous and G. Kuhn, "Induction of finite-state languages using second-order recurrent networks," *Neural Computation*, vol. 4, no. 3, p. 406, 1992.

[15] R. Williams and D. Zipser, "A learning algorithm for continually running fully recurrent neural networks," *Neural Computation*, vol. 1, no. 2, pp. 270–280, 1989.

[16] Z. Zeng, R. Goodman, and P. Smyth, "Learning finite state machines with self-clustering recurrent networks," *Neural Computation*, 1993. To be published.

Implementing the Optimal CDMA Multiuser Detector with Hopfield Neural Networks

George I. Kechriotis and Elias S. Manolakos

Communications and Digital Signal Processing (CDSP)
Center for Research and Graduate Studies
Electrical and Computer Engineering Dept.
Northeastern University, Boston, MA 02115

Abstract

We propose a novel implementation of the CDMA optimal multiuser detector (OMD) based on the Hopfield Neural Networks (HNN). We show how the OMD's objective function can be mapped to a HNN energy function and we evaluate the proposed detector via extensive simulations and theoretical analysis. Its is shown that the HNN detector outperforms by far the conventional detector and has a performance comparable to that of the OMD at a much lower computational cost.

1 Introduction-Background

The optimum centralized demodulation of the information sent simultaneously by several users through a Gaussian multiple access channel is a very important problem arising in multipoint-to-point digital communication networks (e.g radio networks, local-area networks, uplink satellite channels, uplink cellular communications). In the *Code Division Multiple Access* (CDMA) technique, each transmitter modulates a different signature signal waveform, which is known to the receiver. At the receiver, the incoming signal is the sum of the signals transmitted by each individual user.

To demodulate the received signal we need to suppress the inherent channel noise, often modeled as an additive Gaussian process, and multiple access interference (MAI). If the waveforms assigned to each of the users are orthogonal and the transmitted signals are binary ($\{-1, +1\}$) then the *conventional detector* can recover the information by first passing the received process through a bank of filters matched to the users' signature waveforms and then deciding on the information bits based on the sign of the output. One of the major limitations of the conventional detector is that it is not *near-far* resistant, i.e its performance degrades severely when powers of the transmitting users are very dissimilar.

In [1], Verdu showed that in both the synchronous and the asynchronous case, the optimum near-far resistant multi-user demodulation is an *NP-complete* problem that is equivalent to the maximization of an integer quadratic objective function. Therefore, research efforts are concentrated on the development of suboptimal receivers, that are near-far resistant, have low computational complexity, and their performance is comparable with that of the optimum receiver.

Among the suboptimal detectors that have been proposed we mention the *decorrelating detector* by Lupas and Verdu [2], and the *multistage detector* by Varanasi and Aazhang [3]. Recently, Aazhang et all [4], and U. Mitra and H. V. Poor [5], have proposed feedforward neural network based multi-user detectors. While their performance is very good for a small number of synchronous or asynchronous users, their hardware complexity appears to be *exponential* in the number of users, as conjectured in [4]. In this paper we propose an alternative multiuser detector that is based on the ability of the Hopfield Neural Networks (HNN) [6] to give very fast suboptimal solutions to hard combinatorial problems.

2 HNN Formulation of the OMD problem

Assume that K active transmitters share the channel at a given time instance. A signature waveform $s_k(t)$, time limited in the interval $t \in [0, T]$ is assigned to each transmitter. Let's denote the ith information bit of the kth user as $b_k^{(i)} \in \{+1, -1\}$. In a general CDMA system, the signal at a receiver is the superposition of K transmitted signals and additive channel noise :

$$r(t) = \sum_{i=-P}^{P} \sum_{k=1}^{K} b_k^{(i)} s_k(t - iT - \tau_k) + n(t), \quad t \in \mathbf{R} \tag{1}$$

In (1) $\tau_k \in [0, T)$ are the relative time delays between the users and $2P + 1$ is the packet length. In case that the stations cooperate to maintain synchronism, it holds that $\tau_k = 0, \quad k = 1, \ldots, K$. As we mentioned in the previous section, the conventional receiver consists of a bank of filters matched to the signature waveforms of each user that decides on the ith

information bit $\hat{b}_i$ based on the sign of the output of the ith matched filter y_i :

$$\hat{b}_i = sgn(y_i) \tag{2}$$

The *optimal* receiver on the other hand, decides on the information vector $\mathbf{b}_{opt}$ based on the maximization of the log of the likelihood function. In the *synchronous* case it holds that:

$$\mathbf{b}_{opt} = \arg \max_{\mathbf{b} \in \{+1, -1\}^k} \{2\mathbf{y}^T\mathbf{b} - \mathbf{b}^T\mathbf{H}\mathbf{b}\} \tag{3}$$

where $\mathbf{H}$ is the $K \times K$ matrix of signal cross-correlations, and $\mathbf{y}$ consists of the outputs of a bank of K filters, each matched to one of the signals.

It is apparent from Eqn. (3) that the optimal multiuser detector objective function is very similar to a HNN energy function. It has been shown [6], that in the case of symmetric connections ($T_{ij} = T_{ji}$), the equations of motion for the activation of the neurons of a HNN *always* lead to convergence to a *stable state*, in which the output voltages of all the amplifiers remain constant. Also, when the diagonal elements (T_{ii}) are zero and the width of the amplifier gain curve is narrow, (i.e the nonlinear activation function $g(\cdot)$ approaches the binary thresholding function), the stable states of a network of N neuron units are the *local* minima of the quantity (*energy function*) :

$$E = -\frac{1}{2}\sum_{i=1}^{N}\sum_{j=1}^{N} T_{ij}V_iV_j - \sum_{i=1}^{N} V_iI_i \tag{4}$$

With the exception of pathological cases, when the matrix $\mathbf{T}$ is negative or positive definite, networks with vanishing diagonal elements have minima *only* at the *corners* of the N-dimensional hypercube $[0 \;\; 1]^N$ [6].

To use a HNN for solving efficiently the combinatorial optimization problem in Eqn. (3), the quadratic objective function has to be transformed into the energy function of a HNN in the form of Eqn. (4). For the case of antipodal transmitted signals, the vector $\mathbf{b}$ has elements in $\{-1, +1\}$, whereas the V_i elements in the HNN energy function assume values in $\{0, 1\}$. The translation/scaling of the $\mathbf{b}$ vectors will result in an equivalent translation/scaling of the $\mathbf{H}$ and $\mathbf{y}$ parameters, but we need to ensure that the original symmetry of the $\mathbf{H}$ matrix is preserved under this transformation. To further transform the $\mathbf{H}$ matrix into a zero-diagonal matrix, the special structure of the problem is exploited and the methods discussed in [7] are used. The mapping method can be easily extended to include cases of non-binary transmitted signals as well. It can be shown that

$$\max_{\mathbf{b} \in \{+1, -1\}^k} \{2\mathbf{y}^T\mathbf{b} - \mathbf{b}^T\mathbf{H}\mathbf{b}\} \;\Leftrightarrow\; \min_{\mathbf{v} \in \{0, 1\}^k} \{-\mathbf{I}^T\mathbf{v} - \frac{1}{2}\mathbf{v}^T\mathbf{T}\mathbf{v}\} \tag{5}$$

where $\mathbf{T} = -2\widehat{\mathbf{H}}$ and $\mathbf{I} = \mathbf{y} + \widehat{\mathbf{H}}\bar{\mathbf{1}}$, and $\widehat{\mathbf{H}}$ is obtained by setting all the diagonal elements of $\mathbf{H}$ to zero. It is easily verified that our HNN energy function model can be extended to include higher constellations as well. If for example the transmitted symbols b_i are drawn from the set $\mathcal{B} = \{-3, -1, +1, +3\}$ and $K = 3$, then the objective function can be written as:

$$\min_{\hat{\mathbf{v}} \in \{0,\, 1\}^{2K}} \{-(\mathbf{y}^T + (3\bar{\mathbf{1}}^T\mathbf{H} - \bar{\mathbf{1}}^T diag(\mathbf{H})))\mathbf{P}\hat{\mathbf{v}} + \mathbf{v}^T\mathbf{P}^T\widehat{\mathbf{H}}\mathbf{P}\hat{\mathbf{v}}\} \tag{6}$$

where

$$\mathbf{P} = \begin{bmatrix} 2^0 & 2^1 & 0 & 0 & 0 & 0 \\ 0 & 0 & 2^0 & 2^1 & 0 & 0 \\ 0 & 0 & 0 & 0 & 2^0 & 2^1 \end{bmatrix}$$

and $diag(\mathbf{H})$ denotes the diagonal matrix whose diagonal elements are equal to those of the original matrix $\mathbf{H}$. Thus, the HNN suboptimal detector has a hardware complexity (number of units/OP-AMPs) equal to $K\log_2(|\mathcal{B}|)$ where $|\mathcal{B}|$ denotes the cardinality of the constellation set.

3 Simulation Results

In our simulations a number of synchronous users K is transmitting over a Gaussian channel. The Direct-Sequence Spread-Spectrum Binary PSK (DSSP-BPSK) signaling system is used. In DSSP the code waveforms are given by :

$$s_k(t) = A_k\alpha_k(t)\cos(\omega_c t + \theta_k), \quad k = 1, \ldots, K \tag{7}$$

where A_k is the signal amplitude, ω_c is the carrier frequency and $\theta_k \in [0,\ 2\pi)$ is the phase angle. A spreading sequence $\alpha^{(k)} = [\alpha_0^{(k)}, \alpha_1^{(k)}, \ldots \alpha_{N-1}^{(k)}]$, $\alpha_j \in \{+1,\ -1\}$ is assigned to each of the K users, and the time limited code waveform $\alpha_k(t)$ becomes :

$$\alpha_k(t) = \sum_{j=0}^{N-1} \alpha_j^{(k)}\Pi(t - jT_c) \tag{8}$$

where $\Pi(t)$ is the unit rectangular pulse of duration T_c, and $T = NT_c$.

We used a spreading factor equal to: $N = \lceil(log_2(K))\rceil + 1$, and the transmission rate is assumed to be 100Kbits/sec for each synchronous user. The spreading sequence of the kth user has been set to: $1 < k-1 >_2$ where $< k-1 >_2$ denotes the binary representation of the number $k-1$. (for example, if $K = 4$ then the spreading sequences assigned to the users are: $\alpha^{(1)} = (1,-1,-1)$, $\alpha^{(2)} = (1,-1,1)$, $\alpha^{(3)} = (1,1,-1)$ and $\alpha^{(4)} = (1,1,1)$.) The carrier frequency ω_c is set at 6.4340MHz. The channel is being modeled as Additive White Gaussian Noise (AWGN) channel, and the variance of the zero mean, Gaussian noise is σ_n^2. The HNN receiver is initialized to the output of the conventional receiver. For each realization, $K-1$ users have the same energy per bit the Kth user has 10 times the energy of the other users (maximum *near-far* ratio = 10).

The HNN we used had *binary* activation functions and its weights T_{ij} were preset according to Eqn. (5). For each sample of the output of the matched filter bank $\mathbf{y}$, the appropriate biases $\mathbf{I} = \mathbf{y} + \hat{\mathbf{H}}\bar{\mathbf{1}}$ are computed and the HNN is being initialized to the output of the conventional detector and then left to converge to its final stable state.

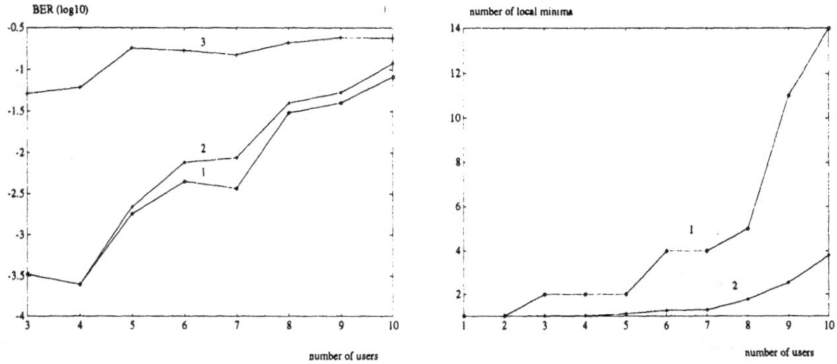

Figure 1: **left**: Bit-Error-Probability (log10 scale) for a number of users ranging from three to ten : (1) optimal receiver, (2) HNN receiver and (3) conventional receiver. **right**: Number of local minima vs. number of synchronous users : (1) maximum, (2) mean.

From Figure 1 (left) where we plot the cummulative Bit-Error-Rate (BER) vs. the number of synchronous users we see that the HNN multiuser detector outperforms substantially the conventional detector and has a performance comparable to that of the OMD. The value value of the SNR was computed with respect to the weak users and the BER was computed from an ensemble of 10^6 transmitted symbols for each value of the SNR. Even for relatively large numbers of active users the performance of the HNN detector is only slightly worse than the OMD.

In Table 1, we evaluate the performance of the proposed HNN detector versus the power of the additive channel noise. As we can see, the performance of the HNN is very close to the performance of the OMD.

The performance of the HNN detector relatively to that of the OMD depends on the average number of local minima of the objective function in Eqn. (5). If this average grows too fast with the size of the problem, then a very poor performance is expected in the demodulation of a large number

SNR	HNN	optimal	conv.	SNR	HNN	optimal	conv.
3.0dB	-1.096	-1.096	-0.935	7.0dB	-1.815	-1.812	-1.104
4.0dB	-1.256	-1.252	-0.990	8.0dB	-2.101	-2.096	-1.146
5.0dB	-1.373	-1.366	-1.017	8.5dB	-2.246	-2.220	-1.184
6.0dB	-1.594	-1.609	-1.068	9.0dB	-3.037	-3.037	-1.204

Table 1: BER (log10 scale) for a four-user case, versus the power of the additive channel noise, evaluated from an ensemble of 10^5 symbols. The SNR is computed with respect to the weak users.

of users.

To experimentally investigate the number of local minima of the objective function (3) we computed using exhaustive search, all the local minima for the number of synchronous users K ranging from two to ten, by applying 1,000 tests for each case. The weak users' SNR was set to 10dB and the maximum near-far ratio was equal to 10 as before. The results are shown in Figure 1 (right). As we can see, the growth of the mean number of local minima is approximately proportional to $K^{\frac{1}{2}}$.

Preliminary results on the conditions on $\mathbf{H}$ and $\mathbf{y}$ that ensure the existence of a unique local minimum have been derived. We have shown that if

$$\sum_{j=1,\ j\neq i}^{K} |h_{ij}| < |y_i|, \quad i = 1, 2, \ldots, K \tag{9}$$

a *unique* local minimum is guaranteed, i.e the HNN is always going to converge to the right solution. It is interesting that these conditions are usually met in practice so that our HNN approach seems to be a viable one. Moreover, the number of local minima can be a-priori determined by inspecting the conditions of Eqn. (9), and for a given cross-correlation matrix $\mathbf{H}$ it depends only on the output of the matched filters $\mathbf{y}$. Given that on the average $\mathbf{y}$ depends on the power of the additive noise and the relative powers of the users, one can analytically determine the long term achievable BER of the HNN detector.

4 Conclusions-Future Research Directions

We have shown how HNNs can be used to efficiently implement the optimal multiuser detector and evaluated the performance of the proposed detectors via extensive simulations. The proposed HNN detector has a hardware complexity (number of neurons) *linear* in the number of users and since it can be implemented in analog hardware, its computational complexity is *constant* irrespectively of the number of users. The fast convergence time (analog HNNs converge within a few RC constants, i.e in the order of a few nanoseconds), makes it an attractive candidate for use in high speed data communications.

Although the results reported here are based on the assumption of a synchronous multiuser scheme, they can be easily extended to the asynchronous case too. The optimum detector for the asynchronous case can be also formulated as a similar integer quadratic optimization problem of much larger size than for the synchronous case. We have performed preliminary simulations for a number of asynchronous users ranging from two to ten, each transmitting 30 bits long data packets. Although the size of the HNN is considerably larger in this case (up to 300 units) surprisingly its performance was found to be similar to the synchronous case. Intuitively this result is due to the sparsity of the HNN weight matrix for the asynchronous case.

References

[1] S. Verdu. Computational Complexity of Optimum Multiuser Detection. *Algorithmica*, 4:303–312, 1989.

[2] R. Lupas and S. Verdu. Near-Far Resistance of Multiuser Detectors in Asynchronous Channels. *IEEE Trans. on Comm.*, 38:496–508, April 1990.

[3] M. K. Varanasi and B. Aazhang. Multistage Detection in Asynchronous Code-Division Multiple Access Communications. *IEEE Trans. on Comm.*, 38:509–519, April 1990.

[4] B.-P. Paris B. Aazhang and G. Orsak. Neural Networks for Multi-user Detection in CDMA Communication. *IEEE Trans. on Comm.*, 40:1212–1222, July 1992.

[5] U. Mitra and H. V. Poor. Adaptive Receiver Algorithms for Near-Far CDMA. In *PIMRC 92*, pages 639–644, October 1992.

[6] J. J. Hopfield. Neurons with Graded Response have Collective Computational Properties like those of Two-State Neurons. *Proc. Natl. Acad. Sci. USA*, 81:3088–3092, 1984.

[7] D. V. Tank and J. J. Hopfield. Simple "Neural" Optimization Networks : An A/D Converter, Signal Decision Circuit, and a Linear Programming Circuit. *IEEE Trans. on Circuits and Systems*, 33:533–541, May 1990.

Scheduling Problems in Radio Networks Using Hopfield Networks

Arun Jagota

Department of Mathematical Sciences, Memphis State University
Memphis, Tennessee 38152 USA
`jagota@cs.buffalo.edu`

Abstract

We address the problems of link scheduling (LSP) and broadcast scheduling (BSP) in radio networks. We reduce LSP to BSP to graph coloring to independent set to a Hopfield network. All local minima of the Hopfield network represent feasible solutions of the original problem (LSP or BSP). We show that steepest descent (SD) dynamics on the Hopfield network, from different initial states, emulates various graph coloring heuristics. We report on preliminary experiments on BSP using SD dynamics on the Hopfield network.

1 Radio Networks and Scheduling Problems

Much of the material in this section is paraphrased explicitly or implicitly from [11]. A *radio network* is a set of n *stations* $v_1, \ldots, v_n$ that communicate by broadcasting and receiving signals. Typical examples include packet radio networks, cellular phone networks, and satellite networks. Every station v_i is assigned a nonnegative number $r(v_i)$ called its *transmission radius* which represents the distance, in any direction, up to which a transmission from v_i can reach. If station v_i has no transmission facilities, then $r(v_i) = 0$. We assume that every station can act as a receiver. $d(v_i, v_j)$ denotes the distance between every pair of distinct stations v_i, v_j. A radio network is called *symmetric* if, for every pair of distinct stations v_i, v_j, if v_j can receive v_i's transmission then v_i can receive v_j's. In this paper, we restrict ourselves to symmetric radio networks.

An n-station symmetric radio network has an undirected graph $G = (V, E)$ underlying it, defined as follows. V is the set of stations. The set of edges is $E = \{\{v_i, v_j\} | d(v_i, v_j) \leq r(v_i)\}$, the pairs of stations that can receive transmissions from each other. This paper concerns itself with two scheduling problems on symmetric radio networks. It turns out that these problems can be conveniently defined as certain coloring problems on the graph underlying a radio network. Two vertices v_i, v_j are called *adjacent* if they are connected by an edge.

Definition—Broadcast Scheduling Problem (BSP).
INPUT: A graph G underlying a symmetric radio network R and a set of colors $C = \{1, 2, \ldots, k\}$
FEASIBLE SOLUTION: To every *vertex* of G assign a color under the constraints:

C1 For every pair v_i, v_j of adjacent vertices, color(v_i) $\neq$color(v_j).

C2 For every triplet of distinct vertices v_i, v_j, v_k, if v_i, v_j are adjacent and v_k, v_j are adjacent, then color(v_i) $\neq$color(v_j).

COST FUNCTION: Number of colors used in a feasible solution.
OBJECTIVE: Find a feasible solution with minimum cost.

Let e_{ij} denote an edge between vertices v_i, v_j. Two edges e_{ij}, e_{kl} are called *incident* if they have a common end-point (vertex).

Definition—Link Scheduling Problem (LSP).
INPUT: A graph G underlying a symmetric radio network R and a set of colors $C = \{1, 2, \ldots, k\}$
FEASIBLE SOLUTION: To every *edge* $e_{ij} = \{v_i, v_j\}$ of G, assign a color under the constraints:

C1 For every pair of edges $e_{ij}, e_{kl} \in E$, if e_{ij} and e_{kl} are incident, then color(e_{ij}) $\neq$color(e_{kl}).

C2 For every triplet of distinct edges e_{ij}, e_{kl}, e_{mn}, if e_{ij} and e_{kl} are incident and e_{kl} and e_{mn} are incident, then color(e_{ij}) $\neq$color(e_{mn}).

COST FUNCTION: Number of colors used in a feasible solution.
OBJECTIVE: Find a feasible solution with minimum cost.

BSP and LSP model broadcast and point-to-point transmissions respectively. C denotes the set of available fixed-length time slots (in the case of Time Division Multiplexing) or frequencies (in the case of Frequency Division Multiplexing). A solution to BSP (LSP) gives the minimum number of time or frequency slots that may be used for broadcast (point-to-point) transmissions, without causing any interference as represented by the constraints C_1 and C_2 of BSP (LSP). For further information, see [11].

BSP and LSP are both NP-hard (see [11]). BSP is NP-hard even when restricted to planar graphs [10]. (Radio networks are typically near-planar graphs.) BSP is NP-hard even to obtain good approximate solutions for [10].

2 Reductions to Independent Set

Here we reduce LSP to BSP to graph coloring (GC) to Independent Set (IS). Figure 1 illustrates these reductions. In Section 3, we map IS onto a Hopfield neural network.

One-to-one. A reduction of an optimization problem A to an optimization problem B is called *one-to-one* if every feasible solution in B corresponds to some feasible solution in A and vice-versa.

Order-preserving. A one-to-one reduction of an optimization problem A to an optimization problem B is called *order-preserving* if, for every pair f_1, f_2 of feasible solutions in A: cost(f_1) "better than" cost(f_2) implies that cost($B(f_1)$) "better than" cost($B(f_2)$), where $B(f_1), B(f_2)$ are feasible solutions in B that correspond to f_1, f_2 respectively and "better than" equals $<$ if the corresponding problem is a minimization one, $>$ otherwise.

Remark. Property One-to-one is a stronger version of P1 in [9]; property Order-preserving is an unambiguous version of P2 in [9].

The *line graph* H of a graph G is a graph with a vertex v_i for every edge e_i of G. v_i, v_j are adjacent in H if and only if the corresponding edges e_i and e_j in G are incident in G.

LSP reduction to BSP. Given an LSP instance $\langle G, C\rangle$ where G is the radio network's underlying graph and C the set of colors, construct a pair $\langle H, C\rangle$ where H is the line graph of G. See Figure 1a,b.

Proposition 1 *The reduction of LSP on $\langle G, C\rangle$ to BSP on $\langle H, C\rangle$ as described above is one-to-one and order-preserving.*

Definition—Graph Coloring (GC).
INPUT: A graph G and a set of colors C.
FEASIBLE SOLUTION: To every *vertex* of G, assign a color under the constraint that for every pair v_i, v_j of adjacent vertices, color(v_i) $\neq$color(v_j).
OBJECTIVE: Find a feasible solution in which minimum number of colors are used.

The *distance-2 graph* G_2 of a graph G is a graph whose vertex-set is the same as that of G. v_i, v_j are adjacent in G_2 if and only if v_i, v_j are adjacent in G or there is a vertex v_k such that v_i, v_k are adjacent in G and v_k, v_j are adjacent in G.

BSP reduction to GC. Given a BSP instance $\langle G, C\rangle$ where G is the radio network's underlying graph and C the set of colors, construct a pair $\langle G_2, C\rangle$ where G_2 is the distance-2 graph of G. See Figure 1b,c.

Proposition 2 *The reduction of BSP on $\langle G, C\rangle$ to GC on $\langle G_2, C\rangle$ as described above is one-to-one and order-preserving.*

An *independent set* in a graph is a set of vertices $I \subseteq V$ in which no pair is adjacent. An independent set I is *maximal* if no strict superset of I is also an independent set.

Definition—Independent Set (IS).
INPUT: A graph $G = (V, E)$.
FEASIBLE SOLUTION: A *maximal* independent set.
OBJECTIVE: Find a feasible solution of maximum size.

Remark. The reason why feasible solutions are restricted to *maximal* independent sets is discussed in Section 3.

The *degree* of a vertex v in a graph G is the number of vertices in G that v is adjacent to. Let $\Delta(G)$ denote the maximum degree of a vertex in G. It is a well-known fact that every GC instance $\langle G, \{1, \ldots, \Delta(G)+1\}\rangle$ has at least one feasible solution. A *clique* in a graph is a set of vertices $C \subseteq V$ in which every pair is adjacent.

GC reduction to IS. Given a GC instance $\langle G, \{1, \ldots, \Delta(G)+1\}\rangle$ where G is an n-vertex graph, construct a graph H containing $n \times (\Delta(G) + 1)$ vertices, partitioned into n disjoint sets $V_1, \ldots, V_n$ respectively. $V_i = \{c_1 v_i, \ldots, c_{\Delta(G)+1} v_i\}$ is the vertex-set of partition i. Vertex $c_l v_i$ in H represents the assignment of color c_l to vertex v_i in G. For $i = 1, \ldots, n$, V_i is made a clique. H may be viewed as composed of $\Delta(G)+1$ horizontal layers. The remaining edges in H are composed of $\Delta(G)+1$ copies of G, one for each layer. That is, if v_i, v_j are adjacent in G, then for all $k = 1, \ldots, \Delta(G)+1$, $c_k v_i, c_k v_j$ are adjacent in H. See Figure 1c,d.

Lemma 1 *Every maximal independent set in H contains n vertices, one from each partition.*

Remark. The proof of this Lemma is also contained in subproofs of Claim 1 on page 7 in [1] and Theorem 4 on page 345 in [9]. The Lemma however is not explicitedly stated there, and this is important for our purposes.

Proposition 3 *The reduction of GC on $\langle G, \{1, \ldots, \Delta(G)+1\}\rangle$ to IS on H as described above is one-to-one.*

Remark. The reduction of GC on $\langle G, \{1, \ldots, \Delta(G)+1\}\rangle$ to IS on H as described above is *not* order-preserving as all feasible solutions of the latter have the same cost. In an attempt to rectify this, [1] reduce GC to the maximum weighted independent set problem (MWIS) such that the optimum solution of MWIS is guaranteed to correspond to an optimum feasible solution, i.e. one with minimum colors, of GC. However we have checked that this reduction is *not* one-to-one, in fact there can be feasible solutions of MWIS (maximal weighted independent sets) that do not correspond to any feasible solution of GC. For this reason we do not employ this reduction in the current paper. A theoretically successful attempt to rectify this is in [9]. They also reduce GC to MWIS but their reduction is different from the one in [1]. Structurally, it is our reduction of GC to IS but with weights on vertices. Theorems 3 and 4 in [9] prove that their reduction is one-to-one and order-preserving. The practical difficulty with their reduction however is that their proofs require extreme large vertex weights. This is a problem, as vertex weights get mapped to unit biases when mapping MWIS to Hopfield networks (see Section 3).

3 Independent Set in a Hopfield Network

The Hopfield Network (HN) [3, 4] is a well-known neural network model. In this paper, we assume that the unit states are in $\{0, 1\}$. Our encoding of IS in HN, shown below, is closely related to our previous work [7, 5] and is equivalent [2] to one that has been implemented in optics [12]. It is also essentially the same as the ones in [1, 9, 13]. Our dynamical algorithms for approximately solving IS are however quite different than the ones in [1, 9, 13] (see Sections 4, 5).

IS reduction to HN. Given a graph G, construct a Hopfield network instance H with n units, unit S_i associated with vertex v_i in G. $w_{ij} = -1$ if vertices v_i, v_j are adjacent in G, $w_{ij} = 0$ otherwise. For all i, choose I_i such that $0 < I_i < 1/2$. See Figure 1d,e.

Proposition 4 *The reduction of IS on G to HN on H as described above is one-to-one and order-preserving.*

4 Steepest Descent Dynamics

Recall that in serial updates, from the set of units whose switch would reduce energy, one is picked and switched. *Steepest descent* (SD) is an instance of serial-update dynamics in which a unit whose switch reduces the energy maximally is picked to switch. If there is a tie then the unit with the *smallest index* is chosen to switch. This lexicographic tie-breaking turns out to be a useful feature on our problems (see Section 5). Let SD(V_0) denote, starting with $V_0 \subseteq V$ as initial state, updating units via steepest descent until HcN converges to a local minimum of E.

SD(V_0) on our IS reduction of a graph G in a Hopfield network of Section 3 emulates a greedy graph algorithm on G [6, 5, 2]. We informally sketch the algorithm here as it will be useful in understanding SD(V_0)'s behavior in Section 5, where we will use it to approximately solve BSP from different initial states V_0. The initial network state is $S = V_0$. First, while the network state S is not a clique, vertices (units) in S are switched OFF, one by one. (In each iteration, the vertex in S that is not-adjacent to the largest number of vertices in S is switched OFF. Ties are broken lexicographically.) Second, while the network state S is not a maximal clique, vertices in $V \backslash S$ are added to S, one by one, while maintaining the invariant that S is a clique.

Remark. We have proposed and evaluated other optimizing dynamics more sophisticated than SD(V_0) on hard optimization problems and they work slightly better [5, 8]. Our preliminary experiments indicate, somewhat surprisingly, that on the problems in this paper, SD(V_0) works best. We omit a discussion on this topic for space reasons but point this fact out to explain why we are only considering the SD dynamics.

5 Experiments and Results

Radio Network Model. Our experiments are based on a random symmetric radio network model, characterized by a quadruplet $\langle n, L, B, r\rangle$. n is the number of stations, each of whose positions is generated uniformly at random, independently of others, in an $L \times B$ rectangle. r is the transmission radius, which is the same for all stations.

5.1 SD Heuristics and Insight

Let G be the graph to be colored and H the graph obtained by an IS reduction of GC on G (see Figure 1c,d). Let $l_i = \{c_i v_1, \ldots, c_i v_n\}$ denote the ith horizontal layer of vertices in H (see Figure 1d). Let $L_i = \cup_{j=1,\ldots,i} l_i$

denote the vertices in the horizontal layers 1 through i. In our experiments, we evaluate the heuristic algorithms SD(L_i), for $i = 0, \ldots, \Delta(G) + 1$. We focus on the boundary cases: L_0, which equals $\emptyset$, and $L_{\Delta(G)+1}$, which equals $V(H)$. We first intuitively describe the behavior of SD(L_i) on H in terms of coloring operations on G, omitting the proofs of these descriptions. Imagine that the vertices of H are laid out on a $n \times (\Delta(G)+1)$ lattice. See Figure 1d. We assume that the vertices of H are in lexicographically increasing order in column-major form on this rectangular lattice.

SD(L_0) This heuristic first colors vertex v_1 with color c_1. (Note lexicographic tie-breaking.) Consider now the state when $v_1, \ldots, v_k$ are colored. It then colors v_{k+1} with the *lexicographically first* color whose assignment does not violate any constraints with the colors on $\{v_1, \ldots, v_k\}$. This is a well-known greedy coloring heuristic.

SD(L_i), $i > 0$ This differs from SD(L_0) in that the initial state, on G, is as if each vertex v_j were assigned an initial set of colors $C_j = \{c_1, \ldots, c_i\}$. SD($L_i$)'s operation on H can be interpreted on G as a form of relaxation labeling: colors in the sets C_j are eliminated, one by one, due to constraints from other sets C_k. After the first 'while' loop of $SD(V_0)$ terminates (see algorithm sketch in Section 4), the state of the algorithm is as follows. For every vertex v_k: either $|C_k| = 1$ or $|C_k| = 0$. If $|C_k| = 1$ then $C_k \subseteq \{c_1, \ldots, c_i\}$ and the color of vertex v_k remains *fixed* for the remainder of the algorithm (second 'while' loop).

5.2 Experiments and Results

Due to space limitations, we restrict ourselves to experiments on approximately solving BSP. (Recall that LSP reduces to BSP so most of the setup for approximately solving LSP remains unchanged.) All our experiments below are on random symmetric radio networks with $L = B = 200, r = 20$.

Here we report on the performance of SD(L_i), for different L_i. These experiments were performed only to study the effect of the initial state on the performance of the SD heuristic, and though rudimentary, offer some insights in this regard. (We have done more extensive experiments with different aims but leave them out of this paper for space reasons.) The upper bound on the number of colors required was $\Delta(G) + 1 = 13.8$. (Using that many colors, any algorithm, including ours, can easily color all vertices.) The number of stations was $n = 100$. The results, averaged over six graphs, were 8.85 colors for SD(L_i) for $i = 0, \ldots, 8$ and 11.5 colors for SD($L_{\Delta(G)+1}$). SD(L_0) does much better than SD($L_{\Delta(G)+1}$) because whereas the former starts from the initial state in which no vertex is colored and from a given state colors the next vertex with the *lexicographically least* non-conflicting color, the latter starts with the initial state in which each vertex is colored with every color and starts eliminating a color from a color set associated with some vertex in no particular order of the vertices or the particular colors, only based on the constraints across color sets. Thus $\mathrm{SD}(L_0)$ is more biased towards including less colors than $\mathrm{SD}(L_{\Delta(G)+1})$.

References

[1] D. Ballard, P. Gardner, and M. Srinivas, "Graph problems and connectionist architectures," tech. rep., Department of Computer Science, University of Rochester, Rochester, NY, 1987.

[2] T. Grossman and A. Jagota, "On the equivalence of two Hopfield-type networks," in *IEEE International Conference on Neural Networks*, pp. 1063–1068, IEEE, 1993.

[3] J. Hopfield, "Neural networks and physical systems with emergent collective computational abilities," *Proceedings of the National Academy of Sciences, USA*, vol. 79, 1982.

[4] J. Hopfield, "Neurons with graded responses have collective computational properties like those of two-state neurons," *Proceedings of the National Academy of Sciences, USA*, vol. 81, 1984.

[5] A. Jagota, "Efficiently approximating Max-Clique in a Hopfield-style network," in *International Joint Conference on Neural Networks*, vol. 2, (New York), pp. 248–253, Baltimore, June, IEEE, June 1992.

[6] A. Jagota, "A Hopfield-style network as a maximal clique graph machine," Tech. Rep. 90-25, Department of Computer Science, SUNY at Buffalo, Buffalo, NY, 1990.

[7] A. Jagota, "A new Hopfield-style network for content-addressable memories," Tech. Rep. 90-02, Department of Computer Science, SUNY at Buffalo, Buffalo, NY, 1990.

[8] A. Jagota, "Optimization by reduction to maximum clique," in *IEEE International Conference on Neural Networks*, (New York), pp. 1526–1531, San Francisco, IEEE, March 1993.

[9] J. Korst and E. Aarts, "Combinatorial optimization on a Boltzmann machine," *Journal of Parallel and Distributed Computing*, vol. 6, pp. 331–357, 1989.

[10] S. Ramanathan and E. Lloyd, "Complexity of certain graph coloring problems with applications to radio networks," 1992. Unpublished.

[11] S. Ramanathan and E. Lloyd, "Scheduling algorithms for multi-hop radio networks," in *SIGCOMM'92 Conference Proceedings – Communications Architectures and Protocols*, (New York), pp. 211–222, Computer Communication Review, 1992.

[12] I. Shariv, T. Grossman, E. Domany, and A. Friesem, "All-optical implementation of the inverted neural network model," in *Optics in Complex Systems*, vol. 1319, SPIE, 1990.

[13] Y. Shrivastava, S. Dasgupta, and S. Reddy, "Neural network solutions to a graph theoretic problem," in *Proceedings of IEEE International Symposium on Circuits and Systems*, (New York), pp. 2528–2531, IEEE, 1990.

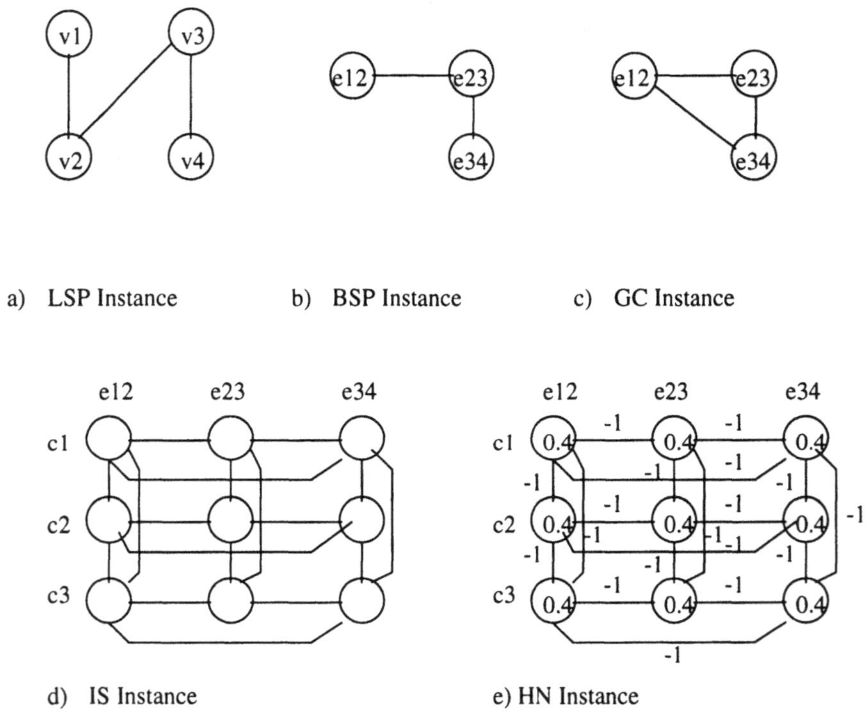

Figure 1: a) is an LSP instance, b) is a)'s reduction to BSP, c) is b)'s reduction to GC, d) is c)'s reduction to IS using $\Delta(G)+1$ colors, where G is the graph in c), and e) is d)'s reduction to a Hopfield network. Edges of G have weight -1, non-edges have weight 0 which is not shown. The numbers in the units are the unit biases (0.4 for all units)

A Hybrid Admission Control Scheme for Broadband ATM Traffic

Ernst Nordström

Department of Computer Systems, Uppsala University
Box 325, S–751 05 Uppsala, SWEDEN
Ernst.Nordstrom@docs.uu.se

Abstract

Link admission control (LAC) in broadband ATM networks is based on evaluation of expected traffic performance. The traditional LAC approach relies on approximate analytical performance models, and can lead to an over controlled network. This paper presents a hybrid LAC scheme which uses a multi layer perceptron (MLP) to refine the performance estimate of a traditional analytical approximation. An accurate but complex traffic performance model is used to derive MLP target data. The numerical results show that the MLP can improve the bandwidth efficiency.

1 Introduction

The Broadband Integrated Services Digital Network (B–ISDN) will be based on a transport mode called Asynchronous Transfer Mode (ATM). ATM is a packet– and connection–oriented switching and multiplexing technique, designed to support virtually all communication services expected in the future. The bandwidth flexibility, the capability to handle all services in an uniform way, and the possible use of statistical multiplexing are advantageous properties of ATM. However, advanced traffic control procedures are needed to maintain the performance of ATM connections. The underlying idea in broadband traffic control is to prevent congestion rather than to react to it. The backbone in ATM traffic control will therefore consist of connection admission control (CAC), traffic enforcement and traffic shaping.

This paper deals with a subroutine of CAC called link admission control (LAC), which is performed at each node along the network path connecting the source and destination node. The purpose of LAC is to establish whether the link can accept a new connection request or if it should be rejected. In particular, we focuses on performance evaluation of a single ATM multiplexer (output buffer in an ATM switch) since this forms the basis for LAC. A trade off between performance evaluation accuracy and computational complexity is necessary to achieve acceptable connection set up delays. The traditional approach is based on analytical approximations, which may result in an over controlled network thereby under utilizing the network resources, see e.g. [3,5,9,11]. This contribution presents a hybrid approach, which use neural networks to improve the link bandwidth efficiency. The suggested scheme comprises a supervised multi layer perceptron (MLP) [4], combined with a tractable analytical approximation. To this end, we employ a variant of the approximate fluid flow performance formula proposed in [8]. This simplified analytical approach will always underestimate link performance. Hence, the MLP's role is to refine the evaluation result, by applying complementary knowledge of the actual performance. Here, the accurate heterogeneous fluid flow model [6,7] and approximations of it, are used to derive additional performance data.

2 Traffic Burst Model

The traffic carried by an individual link are assumed to originate from independent and periodic on/off sources. An periodic on/off source alternates deterministically between burst (on) periods and silent (off) periods with constant characteristic durations. During the burst period, the source transmits packets (called "cells" in ATM terminology), at a constant characteristic cell rate. According to [8], it is widely believed, although not formally proven, that the periodic on/off source represents the "worst case" output of the traffic enforcement function (e.g. implemented by two running windows or two Leaky buckets). Hence, if the connections are represented by periodic on/off sources, the performance analysis will be conservative. The periodic on/off source is characterized by three parameters [8,9]:

- p: the *peak cell rate*, defined as the reciprocal of the constant inter–cell time within a burst,
- m: the *mean cell rate*, defined as the as fraction of time the source is active, multiplied with the peak cell rate, and
- t_{on}: the *burst period duration*

It is very difficult to evaluate the link performance in the deterministic case. However, it is known that more variable burst– and silent–period durations lead to larger queueing build up [8]. In particular, if exponentially distributed durations are used, as in the fluid flow cased below, the performance analysis will again be conservative.

3 Link Performance Evaluation

The total or superposed traffic stream offered to an ATM multiplexer is usually analyzed in two time scales, called the cell scale and the burst scale. The cell scale considers queue build–up (and buffer overflow) at the output buffer due to simultaneous cell arrivals. The burst scale considers queue build–up due to fluctuations in the total arrival *rate*, which may exceed the link capacity during some periods. Throughout this paper we only consider burst scale performance in terms of the cell loss probability, and assume that the buffer is dimensioned to resolve the cell scale conflicts.

3.1 Fluid Flow Cell Loss Probability

The heterogeneous fluid flow queueing model [6,7] is a very accurate tool for evaluating burst scale performance. This model analyses an ATM multiplexer loaded with connections of c different classes, each comprising N_j independent and homogeneous on/off sources (i.e. with equal traffic parameters p_j, m_j and $t_{on(j)}$). The multiplexer is characterized by a buffer of B cells, and an output capacity of C cells/second. The fluid approach models the traffic stream as a continuos (fluid) rate stream. A c–dimensional Markov chain emulates the arrival rate process. The Markov properties implies exponential active– and passive–period durations, which is of practical importance. The underlying Markov state space $\boldsymbol{S}$ is indexed by the possible active–source combinations, i.e. $\boldsymbol{S} = \{\, \boldsymbol{k}=(k_1, ..., k_c) : 0 \le k_j \le N_j \,\}$. The state space comprises a total number of $N_s=(N_1+1)(N_2+1)...(N_c+1)$ states. Hence, the size of the state space increases rapidly (geometrically) with the class sizes N_j.

Two time–independent probability distribution functions are of prime importance in the fluid analysis:

$\pi = \{\pi_{\boldsymbol{k}}\}_{\boldsymbol{k} \in S}$	: the stationary arrival rate distribution,
$\boldsymbol{F}(x) = \{F_{\boldsymbol{k}}(x)\}_{\boldsymbol{k} \in S,\ 0 \le x \le B}$	: the bivariate buffer occupancy distribution.

$\pi_{\boldsymbol{k}}$ is the overall probability that the sources are in state $\boldsymbol{k}$, and is given by the multi–binomial distribution. $F_{\boldsymbol{k}}(x)$ is the probability that the sources are in state $\boldsymbol{k}$, and that the buffer content is less or equal to x. The $F_{\boldsymbol{k}}$'s are determined by solving a set of first order differential equations associated with the Markov states. The standard spectral expansion (eigenvalue–eigenvector) solution yields [6]:

$$\boldsymbol{F}(x) = \sum_{\boldsymbol{n} \in S} a_{\boldsymbol{n}} \exp(z_{\boldsymbol{n}} x)\ \boldsymbol{\phi}_{\boldsymbol{n}}\ , \qquad (1)$$

where $a_{\boldsymbol{n}}$ are coefficients found from boundary conditions, $z_{\boldsymbol{n}}$ are eigenvalues and $\boldsymbol{\phi}_{\boldsymbol{n}}$ are the corresponding (N_s dimensional) eigenvectors. The coefficients $a_{\boldsymbol{n}}$ constitute the main computational burden and dictates the overall complexity of the fluid flow model. The coefficients are calculated by solving a system of N_s coupled linear equations which requires $O(N_s^3)$ numerical operations. However, a simplified computational approach for estimating the coefficients can be found in [1]. The approximation reduces the overall complexity to $O(N_s^2)$.

Once these probability distributions are found, the cell loss probability is easily calculated as the ratio between the loss cell rate and mean arrival cell rate:

$$P_{loss} = \frac{\sum_{\{k \mid k \cdot p > C\}} (k \cdot p - C)\,(\pi_k - F_k(B))}{\sum_{k \in S} (k \cdot p)\pi_k}, \tag{2}$$

where $\cdot$ denotes scalar vector multiplication, and $\boldsymbol{p} = (p_1, ..., p_c)$. The loss rate is obtained by adding the loss rates of all overload states. For the multi–binomial arrival rate distribution, the mean arrival rate simply equals $\boldsymbol{N} \cdot \boldsymbol{m}$, where $N = (N_1, ..., N_c)$ and $\boldsymbol{m} = (m_1, ..., m_c)$.

3.2 Bounds of the Fluid Flow Cell Loss Probability

The fact that the computational complexity of fluid flow model rapidly increases with the size of the state space N_s, severely limits its practical use. In particular, heterogeneous traffic mixes are hardy tractable in real time. Hence, the upper bound of the fluid flow cell loss probability proposed in [8] is therefore of great importance. Although no formal proof is given that validates the approach, it is supported by our extensive numerical tests. The upper bound is approximated by the expression:

$$\overline{P}_{loss} = P_{loss}(0) exp(z_\infty B) \tag{3}$$

The quantity z_∞ represents the asymptotic exponential decay rate as the buffer size approaches infinity, and is given by the largest negative eigenvalue in the spectral expansion (1). The quantity B denotes the actual buffer size. The magnitude $P_{loss}(0)$ is found by substituting the empty buffer boundary condition ($F_k(0)=0$ for $\boldsymbol{k} \cdot \boldsymbol{p} > C$) in expression (2). The resulting expression involves only the stationary arrival rate distribution:

$$P_{loss}(0) = \frac{\sum_{\{k \mid k \cdot p > C\}} (k \cdot p - C)\,\pi_k}{\sum_{k \in S} (k \cdot p)\pi_k}. \tag{4}$$

As for the computational burden, the decay rate z_∞ has only a complexity which increases linearly with the number of classes c [6]. However, the magnitude $P_{loss}(0)$ may require significant computational effort since the complexity behaves like

$O[(c+N)N_S]$, where N_S equals the size of the state space and $N=N_1+....+N_c$ [9]. As already mentioned, N_S will increase rapidly when the number of classes, or the class sizes increase. Fortunately, there exist approximations of $P_{loss}(0)$ with tractable complexity, e.g. Chernoff bound [5,9] or large deviation [5,11]. Both these methods require $O(c)$ operations, and reduce the overall complexity of the final estimate $\hat{\overline{P}}_{loss}$ to the same order.

The hybrid LAC scheme presented in the next section will also make use of a lower bound of the fluid flow cell loss probability, based on the traditional asymptotic fluid flow formula in [7], which states that the spectral expansion (1) can be approximated by the component associated with the largest negative eigenvalue. The lower bound of the cell loss probability is approximated by the expression:

$$\underline{P}_{loss} = -a_{\infty}\, exp(z_{\infty}B)\, \frac{\sum_{\{k \mid k\cdot p > C\}} (k\cdot p - C)\, \phi_{\infty}(k)}{\sum_{k\in S} (k\cdot p)\pi_k}, \qquad (5)$$

where a_{∞} and ϕ_{∞} are the coefficient respectively eigenvector associated with the dominant eigenvalue z_{∞}. The computational approach described in [1] is recommended for estimating the coefficient. In this case, the overall computational complexity of the lower (5) and upper bound (3) become equivalent, i.e. $O[(c+N)N_S]$ in the above notation. However, an efficient lower bound approximation $\hat{\underline{P}}_{loss}$ of the Chernoff/Large deviation type is a subject for future work.

4 Hybrid Link Admission Control

The first objective of link admission control (LAC), is to maintain the link performance, and the second is to optimize the link bandwidth utilization. LAC based on the tractable upper bound yields link performance robustness, but not always link bandwidth efficiency. This section presents an hybrid LAC approach, which combines the tractable upper bound with nonlinear regression (function approximation), to improve this condition. The regression is performed by means of the standard multi layer perceptron (MLP) neural network [4], and concentrates on traffic load situations which are critical in terms of bandwidth utilization.

The hybrid approach is based on the observation that an *upper (lower)* bound of the true cell loss probability always yields correct LAC *accept (reject)* decisions. The inequalities $(\hat{\underline{P}}_{loss} <)\, \underline{P}_{loss} < P_{loss} < \overline{P}_{loss} < \hat{\overline{P}}_{loss}$ is then simply applied in this context. Note that the tractable lower bound is parenthesized since no efficient approximation is known at present, and that the true cell loss probability is represented by the exact fluid flow cell loss probability. The hybrid cell loss probability estimate is defined by a single side inequality:

$$\hat{P}_{loss} = \begin{cases} \bar{P}_{loss} & \text{when } \bar{P}_{loss} < \varepsilon \\ \hat{P}_{loss}(\boldsymbol{L}, \boldsymbol{w}) & \text{when } \bar{P}_{loss} > \varepsilon, \end{cases} \quad (6)$$

where $\hat{P}_{loss}(\boldsymbol{L}, \boldsymbol{w})$ denotes the MLP estimate for input link state vector $\boldsymbol{L}$ and MLP weight vector $\boldsymbol{w}$ (comprising all network weights), and the ε denotes the acceptable cell loss probability (e.g. $\varepsilon=10^{-9}$). According to definition (6), the MLP is only activated when the tractable upper bound states that the performance is unacceptable. The role of the MLP is simply to refine the upper bound estimate, by applying performance knowledge derived by a more accurate/complex analytical model.

The input link state vector $\boldsymbol{L}$ is composed of the tractable upper bound estimate, together with six statistics characterizing the superposed heterogeneous traffic stream [2]. The statistics represent the class parameters $\{N_j, p_j, m_j, t_{on(j)}\}_{j=1,...,c}$ in condensed, approximate form. The most computational demanding statistic is the dominant eigenvalue z_∞. See [2,10] for further details on the preprocessing.

The range of burst level traffic situations considered in the MLP training is derived from a model of offered traffic at the connection level [5]. In the present approach, the offered connection traffic is represented by a traffic intensity function ϱ defined for connection class tuples (N, p, m, t_{on}). In the context of connection level traffic simulation, the class parameters can be seen as stochastic variables with joint distribution function $\varrho(N, p, m, t_{on})$ if ϱ is properly normalized. A traffic situation is simply a set of class tuples drawn from this distribution. Note that the upper bound (3) and lower bound (5) should be consulted before the exact fluid flow model, since one of these bounds could have sufficient accuracy (w.r.t LAC).

5 Numerical Examples

In this section, numerical examples are used to evaluate the accuracy of the hybrid LAC scheme and to compare it to other methods tractable in real time. The Integrated Large Deviation (ILD) [11], Integrated Chernoff Bound (ICB) [9], Equivalent Capacity (EC) [3], and the tractable fluid flow upper bound (implemented by the ILD method) are evaluated separately.

The traffic situations are drawn from independent, uniform and discrete class parameters distributions, where $N \in \{1, 2, ..., 25\}$, $p \in \{2, 4, ..., 14\}$ Mb/s, $m \in \{1, 2, ..., 14\}$ Mb/s and $t_{on} \in \{1, 2, ..., 25\}$ msec. Each traffic situation is composed of three class tuple samples ($c=3$), and fulfill the criterion for statistical multiplexing ($\boldsymbol{N} \cdot \boldsymbol{m} < C < \boldsymbol{N} \cdot \boldsymbol{p}$). The multiplex system is characterized by B=100 cells and C=135 Mb/s, and the LAC by $\varepsilon=10^{-9}$. Given the above constraints, a set of 23512 traffic situations have been generated. According to the exact fluid flow model, 23.2% of these are accept– and 76.8% are reject–LAC situations. The following table shows the adequacy/applicability of the hybrid performance models.

Lower Bound	Exact Fluid Flow	Upper Bound	Tractable Upper Bound
70.8 %	10.4 %	0.001 %	18.7 %

As shown, the lower bound is sufficient (w.r.t. LAC) for many traffic situations. An MLP training set of 20000 samples and test set of 3512 samples are constructed from the generated data. The target performance values are logarithmically transformed to enable representation of tiny probabilities. A weighted MLP error function is used to enhance errors near the LAC decision boundary ε.

The LAC test set results are given as % bad accepts and % bad rejects, which can be interpreted as measures of performance robustness and bandwidth efficiency respectively. The hybrid result is a median over several backpropagation [4] training sessions. As shown in the following table, the hybrid approach yields fewest bad rejects, and therefore has potential for high bandwidth efficiency. Even though it makes some bad accepts, the true cell loss probability is always less than 10^{-8} in these cases.

	Hybrid	Tractable UB	ILD	ICB	EC
Bad Accepts	0.5%	0%	0%	0%	1.1%
Bad Rejects	1.0%	4.3%	13.0%	15.5%	9.1%

6 Summary

In this paper, we have presented and evaluated a hybrid scheme for link admission control (LAC) in broadband ATM networks. The suggested scheme comprises an approximate analytical performance formula, combined with a supervised multi–layer perceptron (MLP). The analytical formula is a computationally efficient variant of an upper bound of the fluid flow cell loss probability proposed in [8]. The MLP is trained using refined performance knowledge, derived (off–line) from different fluid flow performance approximations. An efficient lower bound of the cell loss probability is of interest in the presented scheme, but is part of future work.

The numerical results show that the lower bound of the fluid flow cell loss probability is adequate for many generated traffic situations. The results also show that the hybrid scheme has potential to improve the bandwidth efficiency.

Acknowledgements

The author would like to thank Mats Gustafsson, Olle Gällmo and Lars Asplund, Uppsala University, and Søren Blaabjerg, ELLEMTEL, for stimulating discussions. The financial support from ELLEMTEL, Sweden is also greatly acknowledged.

References

[1] A. Baiocchi, N. Bléfari–Melazzi, A. Roveri, F. Salvatore, "Stochastic Fluid Analysis of an ATM Multiplexer Loaded with Heterogeneous ON–OFF Sources: an Effective Computational Approach", *INFOCOM'92*, Session 3C.3, Florence, Italy, May 1992.

[2] S. Blaabjerg, "On Two Approximations of the Heterogeneous On/ Off Fluid Queue", *COST 224, Final Seminar*, Oct. 1991.

[3] R. Guérin, H. Ahmadi, M. Naghshineh, "Equivalent Capacity and its Application to Bandwidth Allocation in High–Speed Networks", *IEEE JSAC*, vol. 9, no. 7, pp. 968–981, Sep. 1991.

[4] J. Hertz, A. Krogh, R. Palmer, *Introduction to the theory of neural computation*, Addison–Wesley, 1991.

[5] J. Y. Hui, "Resource Allocation for Broadband Networks", *IEEE JSAC*, vol. 6, no. 9, pp. 1598–1608, Dec. 1988.

[6] S. Jacobsen, L. Dittman, "A Fluid Flow Queueing Model for Heterogeneous On/Off Traffic", RACE BLNT Workshop, Munich, July 1990.

[7] L. Kosten, Stochastic Theory of Multi Handling Systems with Groups of Multiple Sources, *Int. symp. on perf. of comp. comm. systems*, IFIP, Zürich, March 1984 (published by Elsevier).

[8] K. Kvols, S. Blaabjerg, "Bounds and Approximations for the Periodic On/off Queue with Applications to ATM Traffic Control", *INFOCOM'92*, Session A.3.1, Florence, Italy, May 1992.

[9] Y. Miyao, "A Call Admission Control Scheme in ATM networks", *ICC'91*, Session 14.5.1, Denver, Colorado, June 1991.

[10] E. Nordström, O. Gällmo, M. Gustafsson & L. Asplund, "Statistical Preprocessing for Service Quality Estimation in a Broadband Network", *WCNN'93*, Portland, Oregon, vol. 1, pp. 295–299, July 1993.

[11] J. Virtamo, "A Note on Hui's Admission Control Algorithm", *COST 224*, TD(90) 038, 1990.

Neural Networks for ATM Multimedia Traffic Prediction

AHMED A. TARRAF, IBRAHIM W. HABIB, and TAREK N. SAADAWI

Electrical Engineering Department
The City College of New York
New York, N.Y., 10031
e-mail: eeaat@ees1s0.engr.ccny.cuny.edu

Abstract

Asynchronous Transfer Mode (ATM) Broadband networks support a wide range of multimedia traffic. Accurate characterization of the multimedia traffic is essential, in ATM networks, in order to develop a robust set of traffic descriptors. Such set is required, by the Usage Parameter Control (UPC) algorithm, for traffic enforcement. It is, also, required, by the Call Admission Control (CAC) algorithm for bandwidth allocation utilizing the statistical multiplexing gain. In this paper, we present a novel approach to characterize and model the multimedia traffic using Neural Networks (NNs). A backpropagation neural network is used to characterize and predict the statistical variations of the packet arrival process resulting from the superposition of N packetized video sources and M packetized voice sources. The accuracy of the results were verified by matching the Index of Dispersion for Counts (IDC), the variance, and the autocorrelation of the arrival process to those of the NN output. The results show that the NNs can be successfully utilized to characterize the complex non-renewal process. Hence, NNs have an excellent potential as traffic descriptors for the (UPC) and the (CAC) algorithms.

1 Neural Network Traffic Predictor

The main objective of this paper is to explore how a three-layered backpropagation NN can be used to characterize and predict non-renewal type processes. NNs based on backpropagation algorithm, can learn a nonlinear relation between many variables. These networks have been used in a number of deterministic and stochastic problems [1], [2]. It has been found that these networks perform well in most cases with accurate results.

Before we start the analysis of the NN model for traffic prediction , we briefly describe the video arrival process and the voice arrival process as follows:

Video Arrival Process: In this paper, the video traffic is generated by simulating a variable bit-rate (VBR) video source which comprise mainly head and shoulder video sequences types without scene changes. The change in the coding bit rate process of a VBR video source is described by a number of mathematical models [3],[4], however, the burstiness of the (VBR) video source depends on the compression algorithm and the nature of the video scene. In this paper, the bit rate encoded information is evaluated frame by frame. Rate variation caused by the video signal line scanning is assumed to be smoothed out before packet assembly. The simulation model used to generate the (VBR) video coded traffic is a continuous-state discrete-time stochastic process. A first order Autoregressive (AR) markov process X(n) that takes into consideration the autocorrelation of the sequence is used. The definition of the AR process is as follows [3]:

$$X(n) = aX(n-1) + bW(n) \tag{1}$$

Where X(n) is the bit-rate during the n th frame, W(n) is a sequence of independent Gaussian random variables where a and b are constants.

Voice Arrival Process: This process possesses correlations among the number of packets arrivals in adjacent time intervals. This complexity is due to the bursty nature of the packet arrival process from single voice source [5]. The ON/OFF periodic process is used to model the voice source, where both ON and OFF time periods are assumed to be exponentially distributed random variables with means $1/\alpha$ and $1/\beta$, respectively. During the ON period a fixed number of packets is generated, each of duration T.

The role of the NN, in this application, is to capture the unknown complex relationship between the past and future values of the traffic. In other word, the NN is employed as an adaptive predictor that learns the stochastic properties of the traffic. The traffic prediction is achieved as the NN learns the actual probability distribution function (PDF) of the traffic. Figure (1) illustrates the basic idea in

training a NN to act as a predictor. The packet arrival process, from the source(s), is represented by the data vector [H(i+m)] which is the NN target output vector. [H(i+m)] provides the NN with the bit-rate information from which the predictions will be made. The NN predicts the bit-rate variations by exploiting the inherent correlations that exist among the arrivals in the packet arrival process. For training purpose, the input data vector [(H(i)] to the NN, is the delayed value of the data vector [H(i+m)]. The NN, then tries to match the target output data vector [H(i+m)] with its predicted output data vector [$\hat{H}$(i+m)] by adjusting its weights. It, then, follows that when the input to the NN bypasses the delay unit, the output vector [$\hat{H}$(i+m)] is a prediction of the values the traffic will have in the future. The delay unit, shown in figure (1), delays its input [H(i+m)] for m time steps. Assuming that the NN requires a negligible amount of time to compute its output from its input, then the NN, after training, provides estimates for the values of traffic [$\hat{H}$(i+m)] m steps in the future. This approach to adaptive prediction rests on the assumption of a parameterized class of models for functional relationship between the current and past values of the traffic and its later values, or equivalently between earlier values of the traffic and its current values.
Figure (2) Shows the backpropagation NN structure used in this paper. The offered traffic to the NN shown in figure (1) is selected to be the arrival process resulting from superposition of N video sources and M voice sources. The NN model used is

$$[\hat{H}(i+m)] = NN_f\{[H(i)], [W]\} \qquad (2)$$

NN_f denotes neural network transfer function, where [W] presents the weight matrices of the hidden and output layers. The vector [H(i)] can be used to present the instantaneous values of bit-rate over the past measurement period (T_m) up to the present instant i. Alternatively, the same vector [H(i)] can be used to represent the count process N(0,t) which measures the number of packet arrivals in time (0,t). In this paper, the count process N(0,t) is used for traffic descriptor in ATM networks due to its robustness against the delay variability of the packets interarrival times. Thus, the vector [H(i)] consists of m samples of bit-rate process or m samples of the count process. These samples are obtained by sampling the arrival process at every sampling period (T_s). [$\hat{H}$(i+m)] is the output vector from the NN, presenting the expected traffic over the next measurement period (T_m), (see [7]). It then follows that,

$$T_m = m * T_s \qquad (3)$$

The Neural Network input traffic pattern [H(i)] is expressed as

$$[H(i)] = [h(i), h(i-1), \ldots\ldots, h(i-m-1)]^t \quad (4)$$

Where h(i) is the value of h(t) at the sampling instant i expressed in packets/sec. The target traffic pattern for the next m measurement intervals [H(i+m)] is expressed as

$$[H(i+m)] = [h(i+m), h(i+m-1), \ldots\ldots, h(i+1)]^t \quad (5)$$

Where m represents the number PEs in the NN input and output layers.
The proposed model was verified by matching the statistical characteristics of the arrival process to those of the proposed NN model. The variance of the number of packet arrivals in an interval (0,t) V(t), the IDC, and the R(n) were used to measure how well the output process, from the NN, matched the actual arrival process. These parameters were chosen since they could best capture the increase in the variance of the arrival process over the sum of consecutive intervals [6].

2 Simulation Results

Extensive simulations were performed to obtain the NN data set, for both training and production phases, and also to assess the performance of various NNs architectures. The packet arrival process, resulting from superposition of N packetized video sources and M packetized voice sources, was simulated on a Sun Sparc station 330 using the C language.
In the simulation of the video source, the first order autoregressive model mentioned in section 1 is used. The parameter used in this model are [23]: a=0.8781, b=0.1108, the mean of W(n) is 4.3 Mbps, and the variance of W(n) is unity.
In the simulation of the voice source, the packet generation process of each voice source is modeled exactly as in [7]. The parameter used in this model are: mean active duration $1/\alpha$=350 msec, mean silence duration $1/\beta$=650 msec, fixed packetization period T=16 msec.
To generate the NN data set, the simulation model were run for 30 min operating time and about 100 million packets were generated. In prediction of the video traffic, the sampling period T_s was chosen to be 1/30 sec. Whereas in prediction of the multimedia traffic we used T_s=10 msec [7].
The NNs were simulated using HNC Inc. EXPLORENET 3001 package [8]. Several experiments were performed using various backpropagation NNs architectures. During the work-course of this paper, four NNs architecture were used, they are called NN-1, NN-2, and NN-3. NN-1 has 1 Processing Element

(PE) in the input layer, two slabs (layers) in the hidden layer (the first slab has 10 PEs, and the second slab has 5 PEs, and 1 PE in the output layer. NN-1 is referred to as (1,{10,5},1), while NN-2 is (5,{10,5},5), and NN-3 is (5,{20,15,10},5). The activation functions of these NNs were selected to be sigmoid where the steepness parameter (a factor determines the non-linearity of the sigmoid function) was selected by trial and error method to obtain good results (in the sense of V(t), IDC, and R(n)).

3 Numerical Results

In this section, we demonstrate the validity of the proposed NN model by comparing the numerical results obtained from this model to those obtained from the simulation model. Three experiments were performed to produce the results.
Experiment (1):
In this experiment, we used N=10, M=0 (ten video sources and no voice sources). NN-1 was used to predict the homogenous superposition video count arrival process over the next frame interval (m=1, T_s=1/30 sec). Figures (3) and (4) show that the predicted traffic, output from NN-1 has the same statistical characteristics as those of the actual input traffic.
Experiment (2):
In this experiment, we used N=1, M=1 (one video source and one voice source). NN-2 was used to predict the multimedia (heterogenous superposition process) count arrival process over the next five sampling period (m=5, T_s =10 msec). Figures (5) and (6) show that NN-2 characterizes and predicts the multimedia traffic with good accuracy.
Experiment (3):
In this experiment, we used N=5, M=5 (five video sources and five voice sources). NN-3 was used to predict the multimedia (heterogenous superposition process) count arrival process over the next five sampling period (m=5, T_s =10 msec). Figures (7) and (8) show that NN-3 predicts the multimedia traffic with reasonable accuracy. By comparing these results with those of experiment (1), we can conclude that the homogenous superposition arrival process (as in experiment (1)) is much simple to be characterized and predicted than the heterogenous one, this is because the heterogenous arrival process is more burst than the homogenous arrival process.

4 Conclusions

Based upon the above experiments, we conclude that a suitable NN architecture can be chosen to characterize a specific type of traffic. After completing the training phase of the NN, it learns the PDF of the offered traffic. Hence, the NN can be used as an effective traffic descriptor. It describes the traffic by its actual

PDF (instead of the approximated simple parameters such as the peak and mean bit-rates). In ATM networks, traffic management techniques require traffic parameters that can capture the various traffic characteristics. Adaptability to changes in the traffic characteristics is also important for robustness. The proposed model using the NNs is suitable for implementing an effective ATM traffic descriptor (UPC), since it can adaptively predict the traffic by learning the relationship between the past and the future traffic variations. The potential of the proposed model is demonstrated by its efficacious to predict the packet arrival process of the superposition of N video sources and M voice sources. The results show that NNs can be trained to learn the PDF of the arrival process hence it can function as an adaptive predictor.

References

[1] R. Lippmann "An introduction to computing with neural nets," IEEE Assp Mag., April 1987.
[2] R. Nielsen "Neurocomputing," Addison wesly publishing comp., 1989.
[3] B. Maglaris et al "Performance models of statistical multiplexing in packet video communications," IEEE Trans. Comm, Vol. 36, No. 7, July 1988.
[4] P. Sen et al "Models for packet switching of variable-bit-rate video sources," IEEE JSAC, Vol. 7, No. 5, June 1989.
[5] J Diagle, and J Langford "Models for analysis packet voice communications systems," IEEE JSAC, Vol. SAC-4, NO. 6, Sept. 1986.
[6] I. Habib, and T. Saadawi "Multimedia Traffic characteristics in broadband networks," IEEE Comm. Mag. July 1992.
[7] A. Tarraf, I. Habib, and T. saadawi "Characterization of packetized voice traffic in ATM networks using neural networks," Globecom'93.
[8] HNC Inc. "HNC ExploreNET release 2.0 operating manual," 1991.

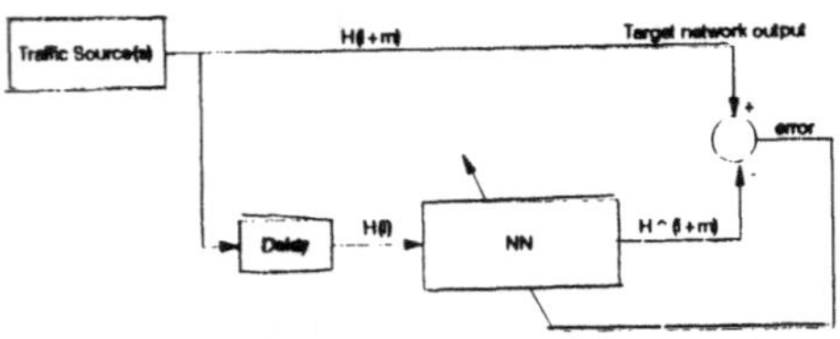

Figure 1: Using Neural Network for adaptive prediction

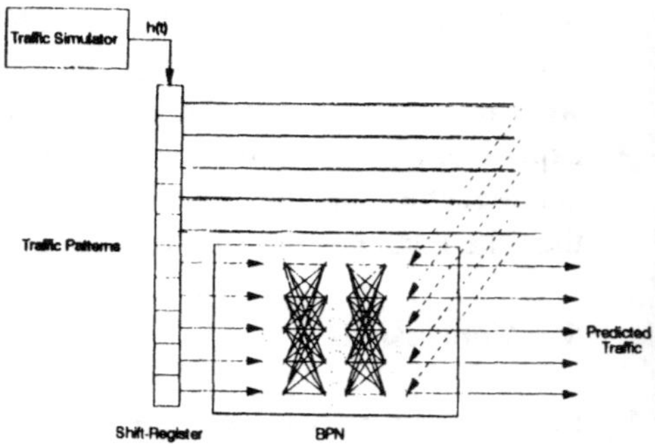

Figure 2: Neural Network for traffic prediction

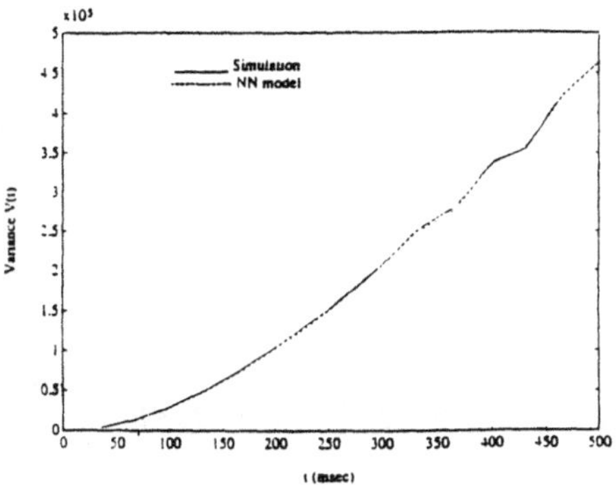

Figure 3: Variance of number of arrivals in (0,t)
(N = 10) (NN-1)

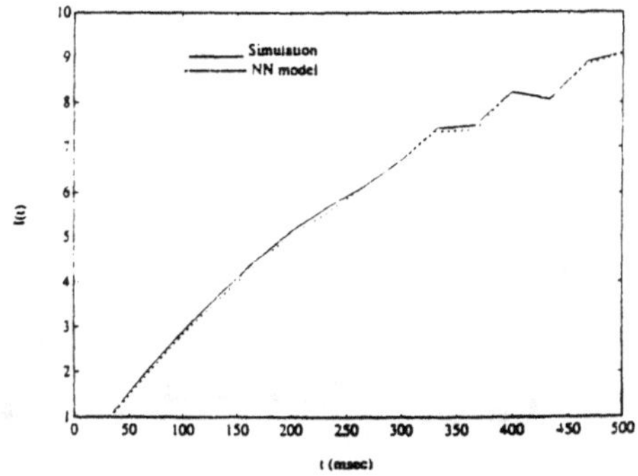

Figure 4: Index of dispersion of number of arrivals in (0,t)
(N = 10) (NN-1)

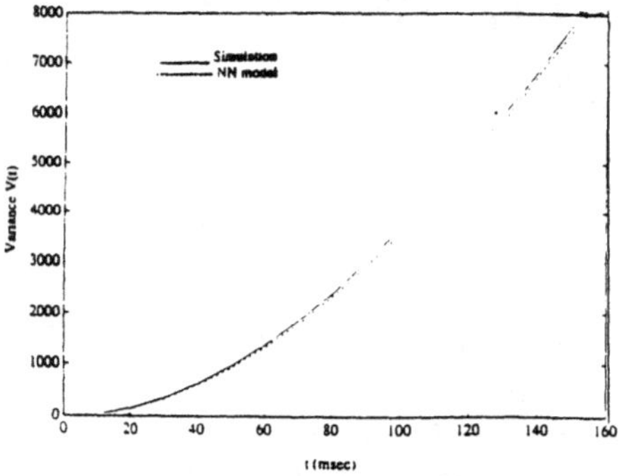

Figure 5: Variance of number of arrivals in (0,t)
(N = 1, M = 1) (NN-2)

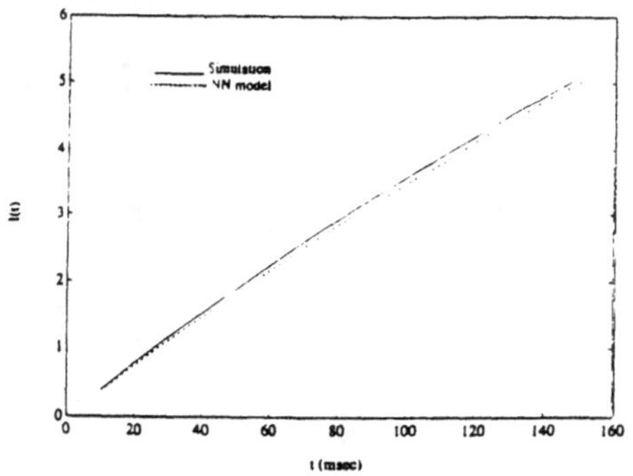

Figure 6: Index of dispersion of number of arrivals in (0,t)
(N = 1, M = 1) (NN-2)

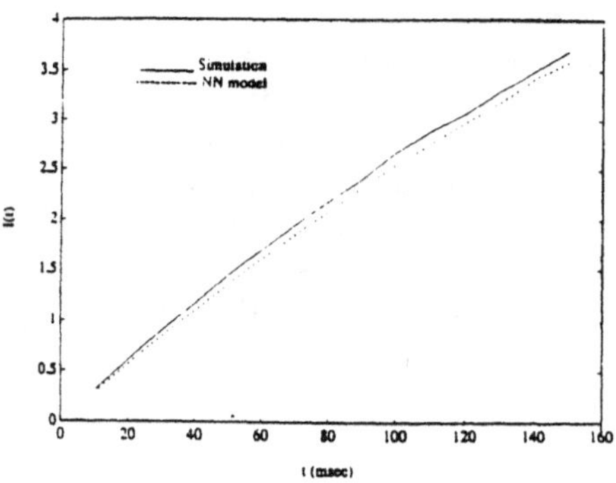

Figure 7: Index of dispersion of number of arrivals in (0,t)
(N = 5, M = 5) (NN-3)

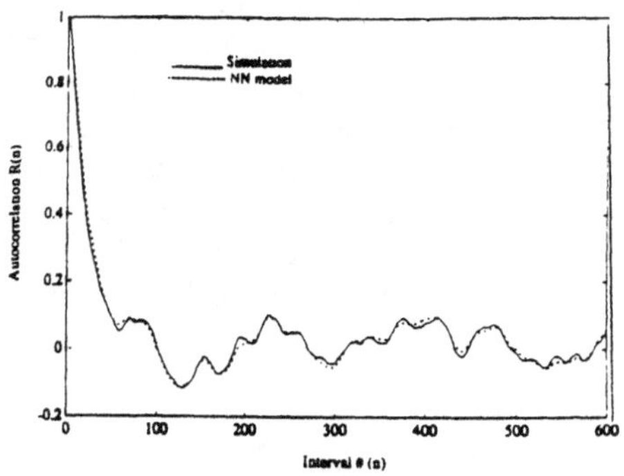

Figure 8: Autocorrelation of the multimedia count process
(N = 5, M = 5) (NN-3)

Table Structure Interpretation and Neural Network Based Text Recognition for Conversion of Telephone Company Tabular Drawings

Atul K. Chhabra, Surekha Chandran, and Rangachar Kasturi*

NYNEX Science & Technology
500 Westchester Avenue, White Plains, NY 10604
atul@nynexst.com
** Department of Computer Science & Engineering*
The Pennsylvania State University, University Park, PA 16802
kasturi@pacific.cse.psu.edu

Abstract

We present a method for extracting structural information of a table and a method for recognizing hand-printed or drafted or machine-printed text from an image of a telephone company tabular drawing. The structure interpretation procedure uses horizontal and vertical lines and white streams as demarcation cues to identify each text block in the table. The character recognition system uses statistical and neural network techniques and a-priori field grammars for interpreting text strings. We are currently working on integrating the two modules.

1 Introduction

Telephone companies have a huge number of records stored in the form of tabular data on engineering drawings. Cable counts, cables in service, assignment charts, cable running lists, and wiring lists are a few examples of such tabular drawing records. This ASCII data has traditionally been stored on paper or microfilm because some of the data dates back to the days when computers were not commonly used in telephone company operations. There is frequent need to access and update this data. Update is performed by highlighting in red and yellow the entries to be added and removed, respectively. The marked and highlighted drawings are then sent to draftspersons for redrawing. The present method of accessing and updating paper drawings is very time consuming and inefficient. Conversion of the drawings to computer data is crucial for efficient operation and for utilizing emerging technologies such as automated mapping and facilities management. Conversion using manual data entry is very tedious and labor intensive. We propose a semi-automatic way of converting the drawings with human supervision needed for correcting any errors made by the system.

In this paper, we describe on-going work in building a system for semi-automatic conversion of the tabular drawings. The pre-cursor to this work-in-progress was the approach presented in [1]. In the work presented here, we use image interpretation techniques for extracting table structure information [2] and a combination of a statistical method and neural networks for text recognition [3]. Currently, we are working on improving the two modules and on integrating them into a complete system.

2 Table Structure Interpretation

Given a binary image of a tabular drawing such as the one shown in Figure 1, the image is deskewed and then scanned to detect all horizontal and vertical demarcation lines. The procedure accepts white streams in horizontal and vertical directions as substitutes for any missing demarcation lines. The extracted demarcation information is used by the structure interpretation procedure to identify each of the text blocks in the table. Interrelations of the text blocks are identified to interpret the structure of the tabulated data. The result of applying this method to the table of Figure 1 is shown in Figures 2–4. The table has all the demarcations present in the form of lines. In addition, there are a few extra vertical lines present in the image. These lines indicate repetition of information across multiple rows. We call these lines ditto lines. A semi-automatic system was used to remove the ditto lines from the list of demarcation features. The final result of structure interpretation of the table is shown pictorially as a tree structure in Figure 4.

This approach is faster and more reliable than the Hough transform technique used in [1]. Also, the approach of [1] only identified text blocks

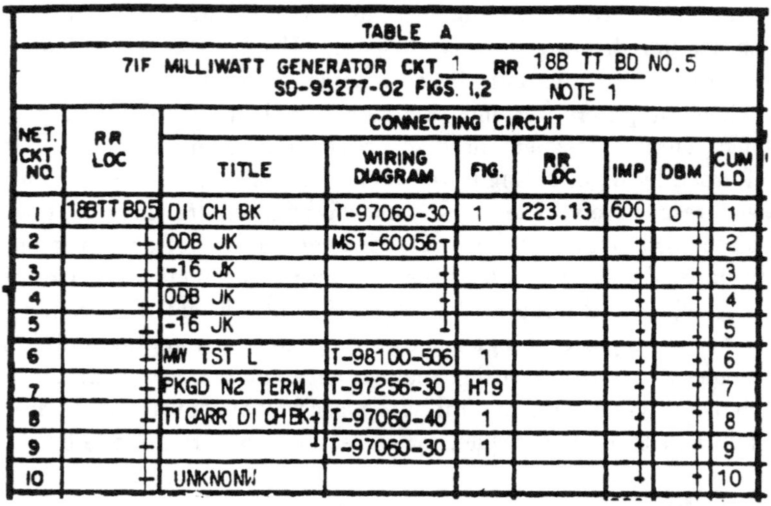

TABLE A								
71F MILLIWATT GENERATOR CKT 1 RR 18B TT BD NO. 5 SD-95277-02 FIGS. 1,2 NOTE 1								
NET. CKT NO.	RR LOC	CONNECTING CIRCUIT						
		TITLE	WIRING DIAGRAM	FIG.	RR LOC	IMP	DBM	CUM LD
1	18BTT BD5	DI CH BK	T-97060-30	1	223.13	600	0	1
2		ODB JK	MST-60056					2
3		-16 JK						3
4		ODB JK						4
5		-16 JK						5
6		MW TST L	T-98100-506	1				6
7		PKGD N2 TERM.	T-97256-30	H19				7
8		T1 CARR DI CH BK	T-97060-40	1				8
9			T-97060-30	1				9
10		UNKNONW						10

Figure 1: Image of a Small Portion of a Telephone Company Tabular Drawing

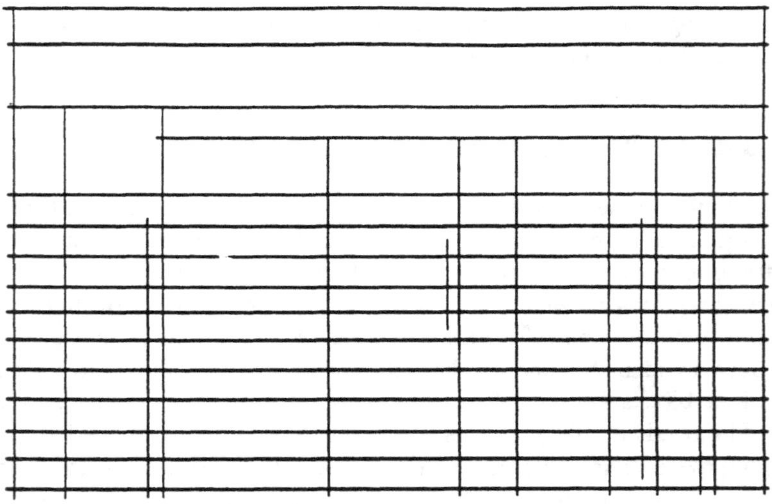

Figure 2: Demarcations Identified in the Drawing of Figure 1

based on demarcating lines. It did not account for missing lines. Moreover, [1] did not build an interpretation of the structure of a table.

We are currently working on automating the detection of ditto lines and the detection of crossed out rows.

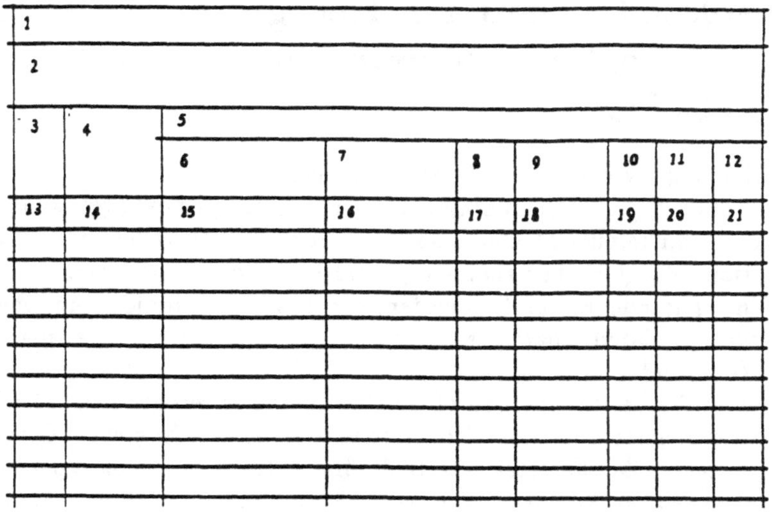

Figure 3: Text Blocks Identified in the Drawing of Figure 1

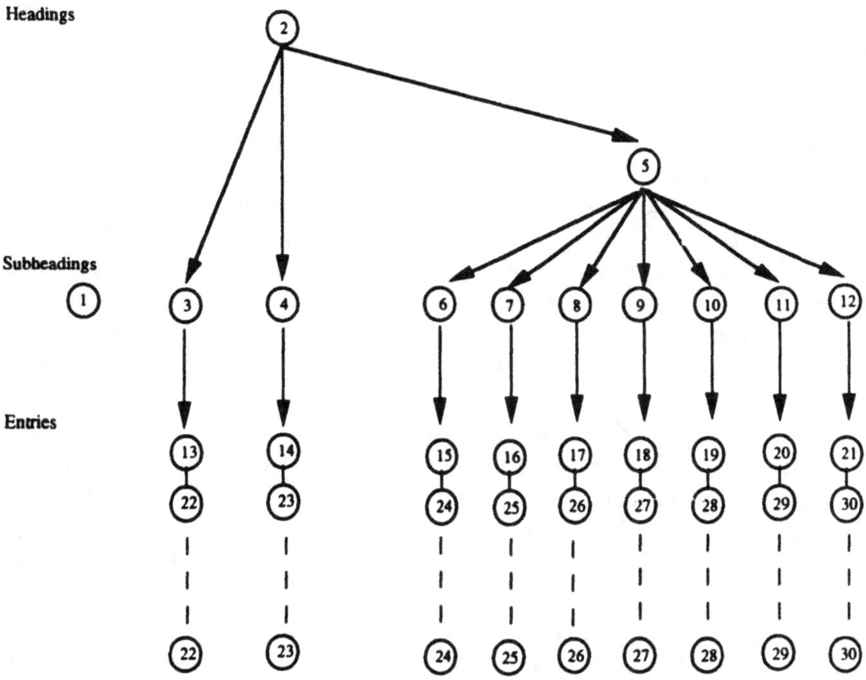

Figure 4: Recognized Table Structure for the Drawing of Figure 1

3 Isolated Character Recognition – Statistical Feature Combination and a Neural Network

The tabular drawings of telephone companies contain text that is hand-printed, drafted (using a template), and/or machine-printed (typewritten). We present an intelligent character recognition (ICR) system for off-line recognition of text strings that are generated by any of the above methods. The character recognition system consists of two major components - the single (isolated) character recognizer (SCR) and the text string segmenter/interpreter.

We first discuss the method of isolated character recognition. The feature extraction stage of the SCR system consists of two steps. Given a bi-level image of a character, the SCR first extracts a set of experimentally optimized raw geometric features. These features are then combined to obtain higher order features which have higher discriminating ability according to the Fisher discriminant measure. The specific combinations of raw features that are chosen in a particular application are determined by statistical analysis of the training data. The combination of raw features introduces a non-linearity in the feature extraction and classification process. This non-linearity is much more powerful than the traditional sigmoid non-linearity introduced by the backpropagation neural network classifiers. The high order features are passed onto a feed forward network (classifier) that is trained using error backpropagation. By using the high order features, we can drastically reduce the number of hidden nodes required in the classifier. In fact, for the ten digit discrimination task, a single layer perceptron performs almost as accurately as a one hidden layer network with 200 hidden nodes. With the feature combiner and a single layer perceptron, the SCR system requires an order of magnitude less number of connections for the same classification accuracy as the backpropagation network. With the raw features and their high order combinations, and given a large training set, we have been able to train a single classifier to accurately read isolated characters from various sources (i.e., hand-printed, drafted, or machine-printed).

Raw geometric feature extraction and high-order feature combination are described in detail in [3]. Here, we present a summary of the procedure. Given a binary image of a character, first the boundary of the character is extracted. The boundary is then convolved with oriented line kernels (horizontal, vertical, and the two diagonals). Also, curvature is computed along the boundary. Convex hull measurements are computed along the boundary. Finally, holes are detected in the character image. The output of each feature computation is then spatially scaled (normalized). This gives us a raw feature vector of 190 floating point values.

From raw features, we compute high-order features which are mathematical combinations of tow or more raw features. High-order features are more powerful discriminators of the character classes. Although there are almost unlimited possible combinations of raw features to make high order features, we limit the choice to products of raw feature pairs. With 190 raw features per image, 18,145 high-order features are possible for each training image. From the 18,145 possible features, we need to select a reasonable subset (e.g. 200 higher order features) to pass on to the classifier. A single raw feature or a pair of raw features is chosen on the basis of how well it separates a pair of classes. Every time we add a feature or a feature pair to our list, we need to ensure that it adds unique information to the features already in the list. This selection of high-order features is accomplished in an off-line iterative process.

Typical methods of character recognition are structural, statistical, and neural-network based methods. At a recent NIST workshop on OCR Systems [4], almost all participants were using neural networks based methods. Neural networks have attracted attention because of their ability to learn a classification task without much manual tweaking and fine tuning. Consequently, most neural network based character recognition systems treat neural networks as a 'black box'. We have shown that significant improvement in classification accuracy and speed can be obtained by intelligent selection of inputs to a neural network classifier [3]. Specifically, we have shown that high order combination of geometric features leads to better accuracy and speed than using pixel data or raw features as input to the neural network.

4 Text String Interpretation

Once a text string has been extracted from the image of a table, and prior to applying the SCR, it is necessary to segment the text string into isolated characters. When neighboring characters are not touching, this is a trivial task. However, in practice, text strings frequently contain characters that are touching each other. We address this problem by initially oversplitting the image of a text string into a large number of segments. These initial segments are heuristically derived using standard rules discussed in character recognition literature. Next, we try to recognize the initial segments and groups of initial segments (grouped using criteria such as proximity) as single characters. We continue regrouping until a reasonable confidence level is attained for each character or until all reasonable groupings have been attempted. In order to better guide the segmentation and grouping process and the overall string interpretation process, we can pre-define grammars for text strings. A grammar defines the structure of a text string. Text entries in tabulated data are often very structured. For instance, text entries in a column of data are likely to conform to similar structure. Grammars can be made more complex by forming conjunctions or disjunctions of several simpler grammars. Thus, the text string interpretation procedure simultaneously performs splitting-regrouping and string-grammar matching.

5 Conclusion

Work is in progress to integrate the table structure interpretation module and the text string interpretation module. Also, we have done preliminary investigation into machine interpretation of schematic (as opposed to tabular) drawings of the telephone company [5].

References

[1] A.K. Chhabra, Z. An, D. Balick, G. Cerf, P. Sheppard, R. Smith, and B. Wittner, "Conversion of Telephone Company Tabular Drawings," in *Proceedings of the International Graphonomics Society's Sixth International Conference on Handwriting and Drawing (ICOHD)*, Paris, France, July 1993.

[2] S. Chandran and R. Kasturi, "Structural Recognition of Tabulated Data," in *Proceedings of the Second International Conference on Document Analysis and Recognition (ICDAR)*, Tsukuba Science City, Japan, October 1993.

[3] A.K. Chhabra, Z. An, D. Balick, G. Cerf, K. Loris, P. Sheppard, R. Smith, and B. Wittner, "High-Order Statistically Derived Combinations of Geometric Features for Handprinted Character Recognition," in *Proceedings of the Second International Conference on Document Analysis and Recognition (ICDAR)*, Tsukuba Science City, Japan, October 1993.

[4] R.A. Wilkinson, et. al., "The First Census Optical Character Recognition System Conference," Technical Report NISTIR 4912, National Institute of Standards and Technology, Gaithersburg, Maryland, USA, May 1992.

[5] J.F. Arias, C.P. Lai, S. Chandran, R. Kasturi, and A. Chhabra, "Interpretation of Telephone System Manhole Drawings," in *Proceedings of the Second International Conference on Document Analysis and Recognition (ICDAR)*, Tsukuba Science City, Japan, October 1993.

Arabic Character Recognition System Using Artificial Neural Network

Adnan Amin and Humoud B. Al-Sadoun*

School of Computer Science and Engineering,
University of New South Wales,
P.O. Box 1, Kensignton,
NSW 2033, AUSTRALIA.
Email: amin@cs.unsw.oz.au
**Department of Electrical and Computer Engineering,*
Kuwait University,
P.O. Box 5969,
Safat 13060, KUWAIT.
Email: hbh@eng.kuniv.edu.kw

Abstract

This paper proposes a structural technique for the recognition of hand(printed) isolated Arabic character using Artificial Neural Networks. Its advantages are: it combines rule-based (structural) and classification tests; efficient for large and complex sets such as Arabic characters; not expensive for feature extraction; execution time does not depend on either the font or the size of the characters.

1 Introduction

Character recognition is an area of pattern recognition that has been undergone a great deal of research during the last twenty years. The recognition and processing of (hand)printed characters is motivated, in large part, by a desire to improve man-machine communication.

To date few products are commercially available to perform such tasks. Furthermore, their abilities are strictly limited to the recognition of typed text. Products to perform hand-printed text recognition are not available, although many techniques have been proposed. Indeed there has recently been a high level of interest in applying artificial neural network architectures to solve the problem [1].

It is well known that the computer recognition of hand-written text is a difficult issue due to the writing style and the character shape variation. The recognition of hand-written text is widely applicable in areas such as: postal code recognition, office automation, and a large variety of other business and scientific applications[2].

The neural networks techniques have been applied to the area of automatic character recognition. Most of the work reported, in this area, during the past decade made use of either pattern matching or statistical approach for features extraction. Pattern Matching is useful for printed characters but will not work for hand-written characters. A number of researchers have recently applied a neural networks technique to recognize the hand-written characters [3-5] by using a statistical approach. However, these have generally been confined to relatively small character sets, usually digits only. The statistical approach is expensive and has not been proven very efficient for large and complex sets.

Arabic, similar to Latin script, is manifest in many fonts and types of writing, e.g., Ruq'a, Nastaliq, Kufi, Royal Diwani, Naskh, etc. Also, some Arabic characters have exactly the same shape, except they are distinguished from each other by the addition of a complimentary character: one, two, or three dots, a zig-zag (Hamza) and/or a vertical bar shape. These complementary characters can be in different positions (above, below, or within the character). Figure 1 depicts some of these characters.

ح خ ج يـ نـ تـ ثـ ئـ ع غ فـ قـ سـ شـ

Figure 1: Arabic Characters Differing only with Regard to the Position and Number of Associated Dots

This paper proposes a structural technique for the recognition of printed and hand-written isolated Arabic characters using neural networks. The advantages of this technique can be summarized in the following points. First, the proposed technique combines rule-based (structural) and classification tests. Second, it is more efficient for large and complex sets, such as Arabic characters. Third, it is not expensive for feature extraction. Fourth, the execution time does not depend on either the font or the size of the characters.

The technique can be divided into three major steps. The first step is the pre-processing step in which the original image is transformed into a binary image utilizing a scanner; thereafter, the binary image is thinned using a parallel thinning algorithm. Second, the skeleton of the image is traced from right to left and a binary tree is constructed. Some primitives, such as: Straight line, Curve, Loop and Double Loops, are extracted form the binary tree. Third, a five layer artificial neural network will be used for the character classification.

2 Pre-Processing

Initially, the document will be scanned, using a 600 dpi commercial scanner. The algorithm implemented in this study is a parallel technique which was presented by Jang and Chin [6]. Figure 2 shows an original scanned image and the resulting skeleton after applying the thinned algorithm.

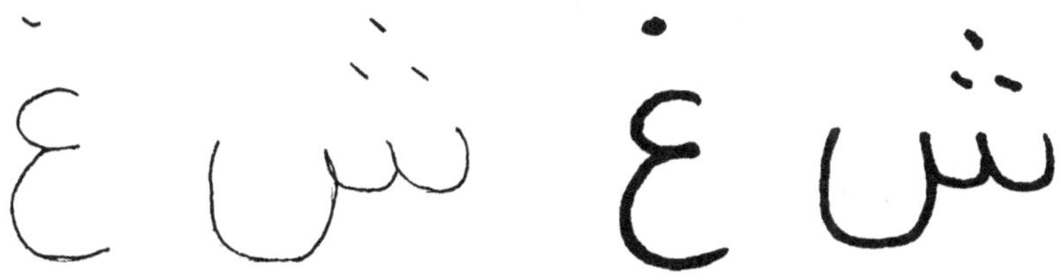

Figure 2: The binary image and the result of the thinning algorithm.

3 Feature Extraction

3.1 Binary Tree Construction

By utilizing a 3x3 grid, the skeleton of the character will be traced from right to left. The Freeman code [7] will be used to describe the skeleton shape. The Freeman code can be described as follows: 0 E, 1 NE, 2 N, 3 NW, 4 W, 5 SW, 6 S, 7 SE. After the construction of the binary tree takes place, the smoothing of the tree will be performed. The objectives of the smoothing of the binary tree are: minimize the number of nodes in the tree;

minimize the Freeman code string; and eliminate any noise in the thinned image.

3.2 Primitives

The algorithm considers two classes of primitives: straight line (S) and curve (C). The straight line has its usual geometrical definition, two points in sequence within the drawing. It is distinguished from a curve in two ways: either directly, if the drawing involves no more than two points; or indirectly, if its length is significant in relation to the rest of the drawing. For straight lines there are eight directions: 0 through 7 (similar to the Freeman code).

Curves are formed by a sequence of at least two segments of nearly equal length (usually small in relation to the total length of the character) with no break point. Two types of curves (Open and Closed (or Loop)) will be considered. For open curves, there are four directions (North, West, East and South). For closed curves (or Loops) there are three sizes (Small, Medium and Large).

At the end of this stage the character should be described in terms of the primitives used in the system.

4 Character Classification

The proposed technique uses a five-layer artificial neural network with one output unit for each possible character output, and with input of a relatively complex nature, described in Figures 3 through 6.

Each character will be classified in terms of the primitives used in the system such as: dot, line, curve, or loop. In each case, the characteristics of each primitive are determined. For instance, in case of the line: its orientation (- | / \ ); its size relative to the character frame (Short, Medium or Long). One input neuron is used to encode each of these possible choices (short/medium/long) and each of four possible orientations for a line. See Figure 3 for details.

The relations between the components are encoded in the *object inter-relationship matrix*, as depicted in Figure 4. This matrix records, for each pair of distinct components, what direction one has to go from the first to reach the second, and also whether the objects touch or not.

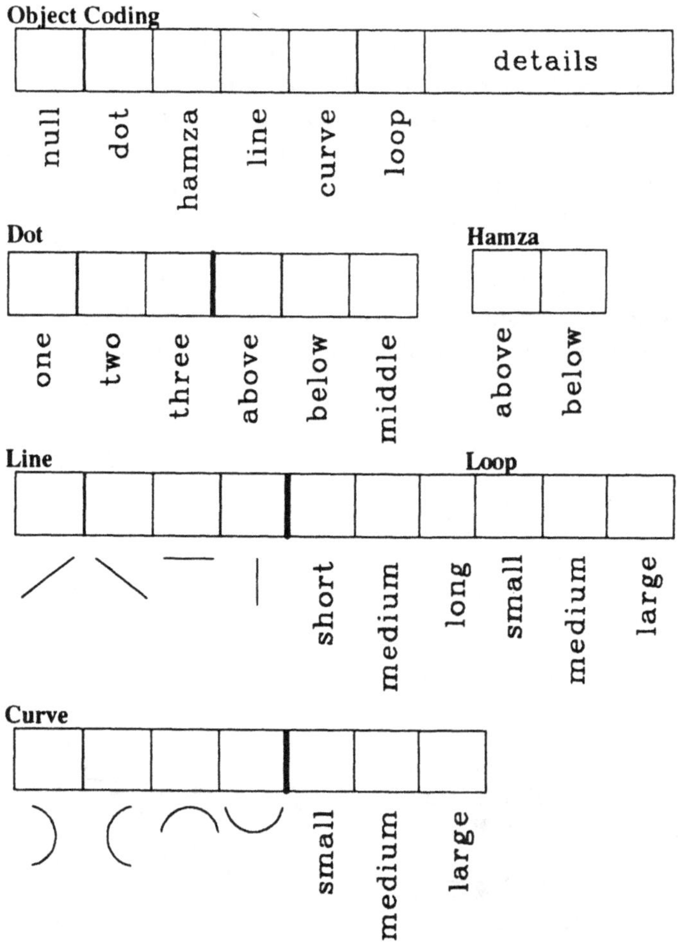

Figure 3: Object encoding for neural network input layer.
Each square cell holds a bit which acts as a flag.

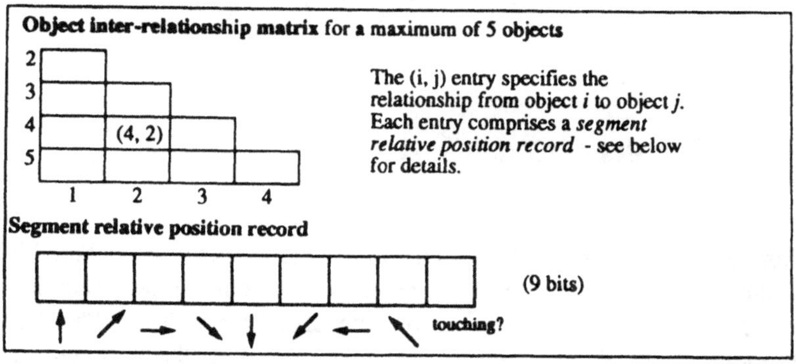

Figure 4: Representation of object interrelationships for neural network input layer.

The overall design of the input layer is shown in Figure 5 for a network capable of handling characters with a maximum of 5 components. It uses a total of 150 input neurons. Figure 6 gives an example of representation of an (imaginary) character using this input layer design. In this example, black squares represent 1 bits, white square represent 0 bits, and gray sections are not used for the particular example.

When the network is trained, the output layer, in response to an input pattern with which it is familiar, or which resembles a familiar pattern, activates the neuron corresponding to this character classification.

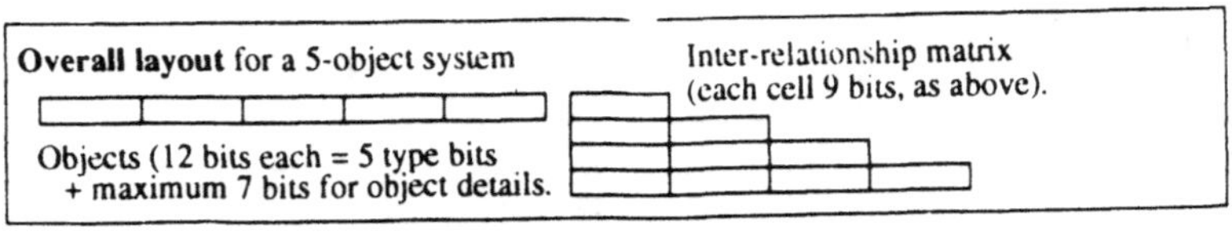

Figure 5: Complete representation of a pre-processed character for neural network input layer.

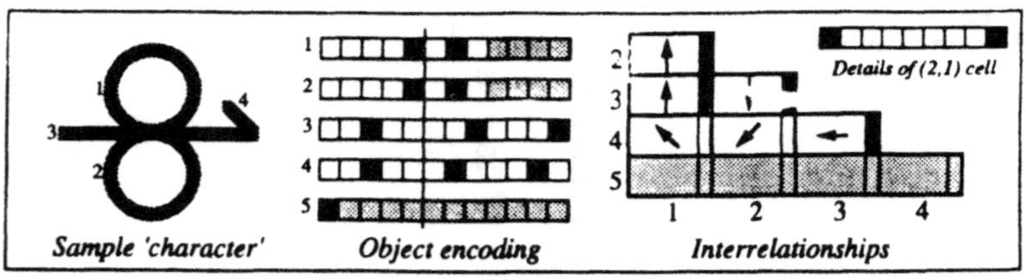

Figure 6: Example for neural network input encoding. Gray fill signifies that the cell is not used.

5 Results & Conclusion

Samples of hand-printed isolated Arabic characters where gathered from a number of persons. The features extracted from some of these samples were sufficient to determine uniquely the character written. The features extracted from the other samples failed to determine the correct identity of the character. However, most of these unrecognized characters where ambiguous to the human eye.

Further work is undergoing to enhance the recognition rate of the isolated characters by incorporating different variants of the shapes of hand-printed characters. Furthermore, the proposed algorithm is being extended to cover hand-(printed) cursive Arabic text.

The study proposed a technique which has been adopted for the recognition of hand-printed isolated Arabic characters using Neural Networks. It combines a purely structural method (based on structural primitives such as curves, straight lines, etc. in a similar manner to which human beings describe characters geometrically) and a classification test.

References

[1] J. Nellis and T.J. Stonham, "An autonomous Digital Neural network Architecture for segmenting Hand-Printed Characters into visually pleasing and low variability sub-classes," *IJCNN-91*, Seattle.

[2] A. Amin and J.F. Mari, "Machine Recognition and Correction of Printed Arabic Text," *IEEE Trans. Sys. Man and Cyber.*, vol. 19, no. 5 pp. 1300-1306, 1989.

[3] P. J. G. Lisboa and S. J. Parantonis, "Invariant Digit Recognition by Zernike Moments and Third Order Neural Networks", in *Proceedings of Second International Conference on Artificial Neural Networks*, 1991, pp. 82-85.

[4] A. Rajavelu and R. Musavi, "A Neural Networks Approach to Character Recognition", *Neural Networks*, vol. 2, pp. 387-394, 1989.

[5] D. Burr, "Neural Net Recognition of Spoken and Written Text," *IEEE Trans. ASSP*, vol. 36, pp. 1162-1168, 1988.

[6] B.K. Jang and R.T. Chin, "One Pass Parallel Thinning: Analysis, Properties and Quantitative Evaluation," *IEEE PAMI*, vol. 14, no. 11, pp. 1129-1140, (1992).

[7] H. Freeman, "On the Encoding of Arbitrary Geometric Configurations," *IEEE Trans. Electronic Comp.*, vol EC-10, pp 260-268, (1968).

Constrained Optimisation for Switching Using Neural Networks

S. Amin, and M. Gell

Systems Research Division, BT Laboratories
Martlesham Heath, Ipswich IP5 7RE, United Kingdom
samin@bt-sys.bt.co.uk

Abstract

A methodical approach to the determination of optimisation parameters which give good convergence in the Hopfield dynamical model for a switching problem is presented. It is shown how the introduction of an imposed attractor into the network enables the model to be computationally operational for all types of random input request matrix.

1 Introduction

A switch can be regarded as a device which takes a set of n input signals and reproduces them in an order required by the switching mechanism. A request for a call connection through that switch can be mapped on to an $n \times n$ binary matrix [1]. In this study we shall be concerned only with square switches of size $n \times n$ .The rows can represent the input lines, and the columns can represent the output lines. During each time slot only one packet can be permitted per channel. Maximisation of the throughput requires effectively an element for continuous optimisation, and it is for this purpose that the neural network is used. A review of applications of neural networks to switching and routing schemes has been given by Brown [1]. In a recent study Ali and Nguyen explored the use of a neural network (the

Hopfield dynamical model [4]) within an access scheme for a high-speed packet switch [2]. Use of a Hopfield neural network for control of a switch was first proposed by Marrakchi and Troudet [3]. Although Ali and Nguyen demonstrated that the Hopfield dynamical network could be used to obtain high solution accuracy, long simulation times were still a problem. This puts severe restrictions on the size of the switching system which can be studied. In this paper a detailed analysis of the optimisation parameters in the Hopfield dynamical equation for the switching problem is presented.

2 The Energy Function

From the analysis of the configuration matrices one can construct an energy function for the switching problem .This function can then be compared with the Hopfield energy function to find the resulting weight connection matrix and the external biases. The energy function for the switching problem is given by [2]

$$E = \frac{A}{2} \sum_{\substack{i,j,l=1 \\ l \neq j}}^{n} y_{ij}\, y_{il} + \frac{B}{2} \sum_{\substack{i,j,k=1 \\ k \neq i}}^{n} y_{ij}\, y_{kj} + \frac{C}{2} \left(\sum_{i,j=1}^{n} n - y_{ij} \right) \tag{1}$$

Now the differential equation for the switching problem can be written as [2], [5],

$$\frac{dx_{ij}}{dt} = -a x_{ij} - A \sum_{l \neq j}^{n} y_{il} - B \sum_{k \neq i} y_{kj} + \frac{C}{2} \tag{2}$$

Under this dynamical equation the neural activities are a dissipative process. They will develop towards neural configurations which minimise the energy function of the switching problem [1] , [5] and are therefore solutions to the switching problem.

3 Establishing the Optimisation Parameters

Now consider the dynamical equation under conditions of equilibrium. From (2) we find

$$x_{0,ij} = -A \sum_{l \neq j}^{n} f(x_{0,il}) - B \sum_{k \neq i} f(x_{0,kj}) + \frac{C}{2} \tag{3}$$

We consider the neural positions ' One ' and ' Zero '. Given that a correct solution to the switching problem has been found this means that the final solution of any input matrix would have at most one active unit per row and per column. First we assume that (i,j) is a zero position.Then the equilibrium condition reads.

$$x^1_{0,ij} = -A - B + \frac{C}{2} \tag{4}$$

In general there will be $n^2 - n$ positions in the network satisfying this condition. This solution may be referred to as the " negative attractor " solution, as it is the equilibrium solution obtained as x tends to minus infinity. The other equilibrium condition is given by

$$x^2_{0,ij} = \frac{C}{2} \tag{5}$$

In general this condition will have to be satisfied at n positions in the network. A symmetric connection matrix can be achieved by putting $A = B$ in which case we find that the parameters have to satisfy the following inequalities

$$0 < C < 4\,A \tag{6}$$

3.1 Imposed attractor

Incomplete requests have to be dealt with carefully so as to make sure that connections are not generated where there are no requests. The switching energy function [2] , [5] does not guarantee this to be the case. This follows from the fact that the switching energy function does not take on minima if one or more rows or columns have only vanishing entries. Dissipative dynamics therefore generate non-zero entries in rows and/or columns which initially have none. To avoid this occurrence, the null columns and rows are decoupled from the Hopfield dynamics. For these special cases, the dynamics has to be modified. This is done in the following manner. All non-zero input neurons will be allowed to change as dictated by the dynamics, whilst all the others will be forced into the negative attractor (the imposed attractor), and all inputs to these neurons are disconnected from all other neural positions in the network. Similarly all the external biases of the zero neurons are set to zero. This arrangement guarantees that these zero neurons stay at their zero output values as required. Without disconnecting the neurons with zero request in the manner discussed above, the network will connect them in order to minimize the energy function.

4 Simulation Results

In all these simulations we introduce noise into the system to avoid spurious convergences. We achieve this by randomly varying the value of β in the activation function [5]. This is to allow the network to escape from trapping minima and to converge to a valid solution. At any fixed time all the neurons will change according to one particular sigmoid function. Extensive simulation have shown that at the limit the magnitudes of $x^1_{0,ij}$ and $x^2_{0,ij}$ are equal, and by putting $x^1_{0,ij} = -\frac{C}{2}$ into equation (4) gives modified versions of equations (6), namely

$$0 < C < 2\,A \tag{7}$$

In the simulation of the Hopfield model for the switching problem we find that any values of A and C which satisfy the above equation give valid results provided that β and δt are correctly chosen. This has been confirmed for a large number of input matrices and different matrix sizes. A practical point is that A is always taken to be very much larger than C. This is due to the fact that a large proportion of the neurons have to approach the negative attractor while only a small number of them will approach the positive attractor. Taking A much greater than C speeds convergence, and allows us to use a reasonably large value of δt. Consider the full 6×6 input request matrix. To understand the behavior of the network as a function of the A, B and C parameters we will study a complete request matrix.

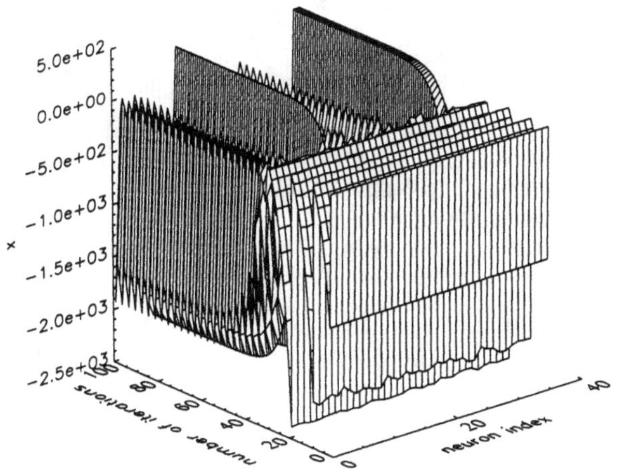

Figure 1: Result for simulations of 6×6 input matrix with $A = B = 1250$ and $C = 700$. Convergence is not reached.

The 3D graphs of the neural input x as a function of time are shown in

Figure 1 and Figure 2 for two different sets of A, B and C.The function in Figure 1 is unstable. Although few neurons are stable, there are a few which are not stable and no solution is achieved. In contrast, the function in Figure 2, attains stability after only 40 time intervals and the system converges to an acceptable solution. It is possible to study the optimisation process for the entire ensemble of 2^{n^2} input request matrices. In the case of $n = 4$, for example, there are 65536 possible requests matrices. Using parameter settings $A = 1250, B = 1250$ and $C = 100$, simulation of the neural network has been performed for each of the ensemble of input request matrices for the cases of $n = 2, 3$ and 4. On average, the computation reaches convergence in less than 20 time intervals for all requested input matrices. More importantly, the neural network finds valid solutions in all cases. In order to test the performance of the neural network for larger networks ($n = 5$ and 6), 100000 input matrices were randomly chosen from the 2^{n^2} possible matrices. Results of the simulations were impressive with convergence being achieved in typically less than 30 time intervals and with all solutions being valid.

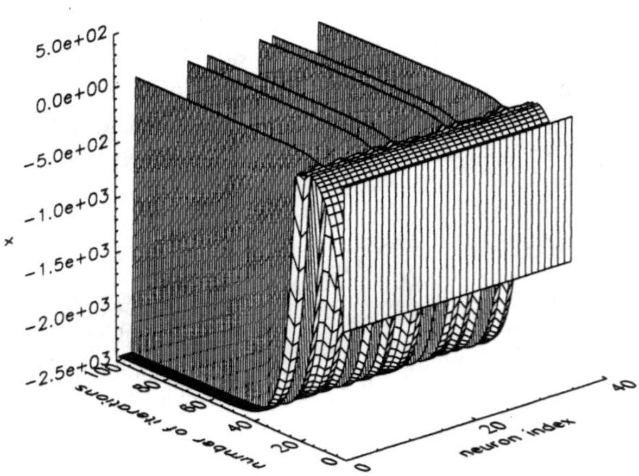

Figure 2: Result for simulations of 6×6 input matrix with $A = B = 1250$ and $C = 100$. Notice that the network convergence after about 40 time intervals

5 Discussion

In this paper the dynamical equation for the switching problem has been assessed with respect to the setting of internal optimisation parameters. By establishing a methodical framework in which values of optimisation parameters (A, B, C) are set and demonstrating the role of an additional imposed attractor $(x^3 = -2\ A)$, it has become possible to carry out simulations for large networks. This enhancement in simulation capability is reflected not only in the establishment of a framework within which valid solutions can always be achieved with large switch sizes but also in enabling those solutions to be achieved with significant computational speed-up. The use of the random gain parameter β is crucial to push the network out of local minima in which the system may get trapped. By choosing the correct values for the parameters A, B and C, we are able to choose a relatively large value for the time step, which improves the speed of convergence.

Acknowledgment

The authors wish to thank Dr J. L. Fernández for helping in the preparation of the manuscript.

References

[1] T.X. Brown, "Neural Networks for Switching," *Special Issue on Neural Networks in Communications,* E.C. Posner, guest editor, IEEE Communications Magazine, p. 72, November, 1989.

[2] M. M. Ali and H. T. Nguyen, "'A Neural Network Implementation of an Input Access Scheme in a High-Speed Packet Switch," in *Proceedings of IEEE Globecom,* 1989, p. 1192–1196

[3] A. Marrakchi and T. Troudet, "A Neural Network Arbitrator for Large Crossbar Packet Switches," *IEEE Transactions on Circuits and Systems,* vol. 36, No 7, p. 1039, 1989.

[4] J. J. Hopfield, "Neurons with graded response have collective computational properties like those of two-state neurons" *Proc. Natl. Sci. , Biophysics, USA* , vol. 81, pp. 3088–3092, 1984.

[5] S. Amin, S Olafsson, and M. Gell, "Constrained Optimisation for Switching using Neural Networks", *European Transactions on Telecommunications* , to be published 1993.

ATM Cell Scheduling for Broadband Switching Systems by Neural Network

Young-Keun Park, Vladimir Cherkassky, and Gyungho Lee

Department of Electrical Engineering,
University of Minnesota, Minneapolis, MN 55455
park@ee.umn.edu, cherkass@ee.umn.edu, ghlee@ee.umn.edu

Abstract

This paper presents a new window-based cell scheduling scheme for a nonblocking ATM switch with input queues, and suggests its neural network implementation. The proposed approach provides an optimal scheduling of ATM cells, in order to achieve high throughput.

1 Introduction

The asynchronous transfer mode (ATM) is being developed as a key technology for the B-ISDN. The ATM is a packet-oriented transfer mode, in which transmitted data are segmented into a number of short fixed-size packets or *cells*. For ATM networks, the *nonblocking switches* such as crossbar switches, rearrangeable switching networks and Batcher-banyan switches have been extensively studied in the literature [1]. In these switches, queueing is required when more than one cell have the same destination address. Based on the placement of buffers, queueing methods can be categorized as *input queueing* and *output queueing*. Even though with output queueing a higher throughput can be achieved as compared with input queueing, the switch must operate N times faster than the channel port speed [3]. In input queueing, the queueing structure is relatively simple but the *head-of-line* (HOL) *blocking* is the major factor of limiting the switch throughput [3]. The HOL blocking is a phenomenon in which a packet in an input buffer has no chance to access an available output port because the packet ahead of it in the buffer is blocked for access to its output port. However, the throughput can be improved if the strict FIFO input queueing is relaxed. Such an approach is known as the *input window policy* [3]. In this scheme, the contention resolution process during a *time slot* is divided into W steps, where W is the window size (see Fig. 1). The first W cells in each input buffer sequentially contend for access to the switch outputs. It is shown that the switch throughput is monotonically increasing with W [3].

2 Proposed Scheduling Algorithm and Neural Network Model

This section describes a new window-based contention resolution algorithm and its neural network implementation. Assume that the switching network is a nonblocking switch that operates with fixed-length cells in a time-slotted fashion. In the proposed algorithm, the scheduling of cells in the window is performed more efficiently and more flexibly than in the original input window policy [3], with an objective of maximizing the number of scheduled cells during a time slot. At every time slot, *all cells* inside the window can contend for their destinations, and an optimal set of nonblocking cells is chosen, under the constraint that *out-of-sequence* transmission should be avoided (see Fig. 2). The following scheduling constraints on the cells in the window must be satisfied at *every* time slot :

1. In each row of the window, at most one cell can be selected to transmit.
2. Every selected cell must have a distinct destination address.
3. In each row of the window, cells with the same destination address must be transmitted in-sequence.

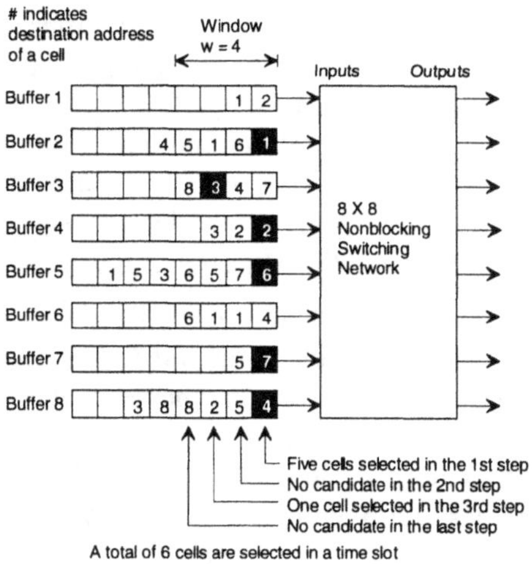

Fig. 1. Original Input Window Policy

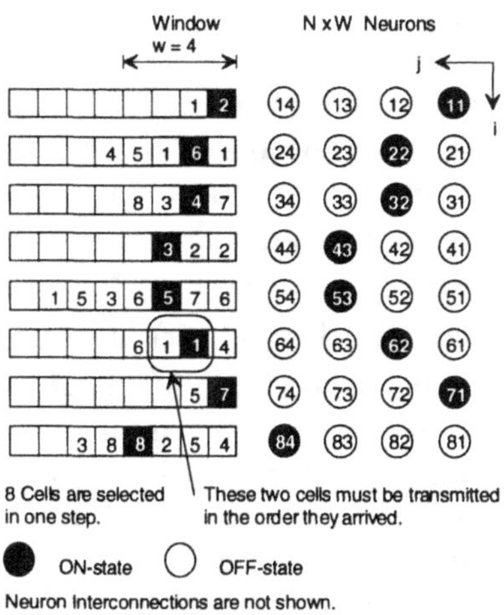

Fig. 2. Proposed Neural Net Scheduling

The objective of scheduling is to maximize the total number of cell transmissions in a time slot.

We use continuous Hopfield-type neural network to solve the above constraint satisfaction problem [6]. The relationship between the input and the output of a neuron is characterized by a *sigmoid function* :

$$V_{ij} = 1/\{\,1 + \exp(-gU_{ij})\,\}, \tag{1}$$

where U_{ij} is the net input to neuron *ij*, V_{ij} is the output, and g is the gain parameter. We define the two active states of a neuron as follows. If the output of a neuron is less than a certain value $\alpha(0 \le \alpha \le 0.5)$, the neuron is said to be in the OFF-state; if it exceeds $(1-\alpha)$, the neuron is in the ON-state; otherwise the neuron is said to be in the intermediate state. The value of α is chosen empirically.

The proposed neural network has *NW* neurons in a two-dimensional structure (Fig. 2). Each neuron corresponds to a cell in the window. If a neuron is ON, the corresponding cell is scheduled to transmit during the current time slot. Connection weights are determined during *every* time slot. Each neuron has two indices *i* and *j*, which represent the input buffer *i* and the position *j* of a cell in buffer *i*. From the three constraints, the following energy function is formulated :

$$E = \frac{A}{2}\sum_{i=1}^{N}\sum_{j=1}^{W}(I_{ij} - \sum_{q=1}^{W}V_{iq})^2 + \frac{B}{2}\sum_{i=1}^{N}\sum_{j=1}^{W}\sum_{\substack{p=1,\\ p\neq i}}^{N}\sum_{q=1}^{W}H_{ij,pq}V_{ij}V_{pq} - C\sum_{i=1}^{N}\sum_{j=1}^{W}(D-T_{ij})V_{ij}, \tag{2}$$

where A, B, C, D : positive constants.

Neuron *ij* has an external input I_{ij}, where I_{ij} is 1 if the *j*th position of buffer *i* contains a cell, and 0 otherwise. This external input forces the neuron *ij* to turn ON if the corresponding position in the input buffer is not empty. Each term in (2) corresponds to a scheduling constraint. The first term is minimized if and only if at most one neuron is ON in each row. In the second term, we define a connection matrix **H**, whose dimension is $NW \times NW$. An element $H_{ij,pq}$ represents the connectivity between the neuron *ij* and the neuron *pq*. If the cell *ij* and the cell *pq* $(p \neq i)$ have the same destination address, the element $H_{ij,pq}$ is 1, and 0 otherwise. Matrix **H** is sparse and it depends on the traffic pattern. During every time slot, the matrix **H** is constructed by observing the destination addresses of the cells in the window. This can be done by the hardware comparators using only XOR gates. The second term in (2) is zero if and only if there is no blocked schedule. The last term enforces the in-sequence transmission constraint by making the internal *thresholds* dependent on the neuron position in the window. Let T_{ij} denote the threshold of neuron ij, then T_{ij} is defined as

$$T_{ij} = \frac{j}{W} \ , \ (1 \le j \le W). \tag{3}$$

Since it is a function of the column index j, the neurons in the same column have the same threshold and the neurons in the rightmost column have the smallest thresholds. The last term in (2) has the minimum value when sum of the thresholds of all ON-state neurons is minimized. In addition to the thresholds, we add a *bias* term D for faster convergence [5]. Inclusions of such terms play an important role in the proposed model, and it is largely a matter of experience. It can be shown that $dE/dt \le 0$, hence, our neural network model always converges to a stable state.

The following motion equation of the proposed model is obtained by taking derivative of the energy equation with respect to V_{ij} (i.e., $dU_{ij}/dt = -\,\partial E/\partial V_{ij}$):

$$\frac{dU_{ij}}{dt} = A(I_{ij} - \sum_{q=1}^{W} V_{iq}) - B \sum_{\substack{p=1,q=1 \\ p \ne i}}^{N \ \ W} H_{ij,pq} V_{pq} + C(D - T_{ij}) \ , \tag{4}$$

In the above equation, the A-term and the B-term imply the inhibitory connections between the neurons. The C-term differentiates between the candidate neurons that have identical values of A-term and B-term. In Fig. 2, for example, neuron 62 and neuron 63 have identical values of A-term and B-term since they have the same source/destination address but neuron 62 will win the competition since it has a lower threshold. Neuron 63 will be selected in a later time slot. Therefore, corresponding cells are transmitted in sequence. The motion equation in (4) forces the neural network into a stable state corresponding to a local minimum of the energy function described in (2). Then the states of neurons directly represent an optimal (or near optimal) schedule of the cells in the window for a time slot.

Since the window size W is fixed (typically less than 8) regardless of the switch size N, the hardware complexity of our model is of $O(N)$ neurons and $O(N^2)$ connections. In the operation of the switch, time slots are established of sufficient duration to allow the cell scheduler to choose the cells for transmission. For a port speed of 500 Mbits/sec, the time slot duration will be around 1 μsec (i.e., the time required to transmit a 53 byte cell in a time slot). Processing time of the proposed approach mainly depends on the sum of the weight loading time and the neural network convergence time. Each connection weight can be stored as a charge on a capacitor and parallel loading of all weights which is widely used in *cellular neural networks* for image processing takes about 1 μsec [9]. In 1989, a Hopfield network with 64 neurons was implemented in a CMOS VLSI, and the convergence time was 100 nsec [8]. Therefore, for a moderate size switch, say up to 128×128, the proposed neural network controller (1024 neurons for $N=128$, $W=8$) seems feasible with the current technology.

3 Empirical Performance Evaluation

In our simulations, we used $A=B=5.0$, $C=2.5$, $D=1.4$ for $W \geq 2$, $D=0.7$ for $W=1, \alpha=0.2$, and $g=2.0$. The *throughput* is defined as the average utilization of the output trunks of a switch. The *maximum throughput* is the throughput for an *arrival rate* equal to 1. The *cell loss probability* is defined as the ratio of the number of buffer overflows, which causes a cell loss, to the number of cells arrived in a given time. The *mean cell waiting time* is the time that a cell is in the buffer until it is transmitted to its output port.

In our simulations we assumed the uniform traffic, and a simulation run comprised 10^4 cell arrivals at a chosen arrival rate λ and a window size W. There was no convergence failure with an update limit of 500, and there was no cell scheduling failure (i.e., blocking or out-of-sequence). The network converged usually within 150 updates per neuron. The maximum throughputs of the proposed model are plotted in Fig. 3(a), and those of the original input window policy are also provided in Fig. 3(b) for comparison. The maximum throughput of our model for $W=8$ is about 0.954. Fig. 4 shows the throughputs of an 128×128 switch as a function of arrival rate for various window sizes. Notice that a window size of $W=1$ corresponds to the strict FIFO queueing discipline and its throughput is saturated at 0.58 in good agreement with the analytic result 0.586 [4].

The cell loss probabilities for every buffer size were estimated from queue length distributions shown in Fig. 5. It represents the probability that k number of cells are in an input queue. As the arrival rate increases, it has a wider distribution. By extrapolating the trails of these curves, we can estimate the cell loss probabilities as shown in Fig. 6 and Fig. 7. We used the loglinear regression method to find estimates of cell loss probabilities. The correlation coefficients were about –0.99, which means that the regression lines fit quite well into the data. As one can see from the figures, the proposed cell scheduling scheme provides lower cell loss probabilities than other approaches.

It can be shown that the proposed input scheduling provides almost the same average queue length and the mean cell waiting time as the output queueing. In high speed networks, propagation delay dominates over queueing delay. A cell loss requires a retransmission, which results in an additional propagation delay. This delay is much larger than queueing delays. Therefore, guaranteeing low cell loss probability is more important than reducing queueing delay in order to meet the QOS requirements.

4 Conclusion

In this paper, we proposed a new window-based scheduling algorithm for input queueing and presented a neural network controller for optimal scheduling. The performance of the proposed approach is also compared with the original input queueing [3] and output queueing [2]. With $O(N)$ number of neurons, the proposed neural network significantly increases the maximum throughput of the switches. The cell loss probabilities and the buffer sizes necessary to guarantee a given cell loss rate become smaller than with other approaches [2,3].

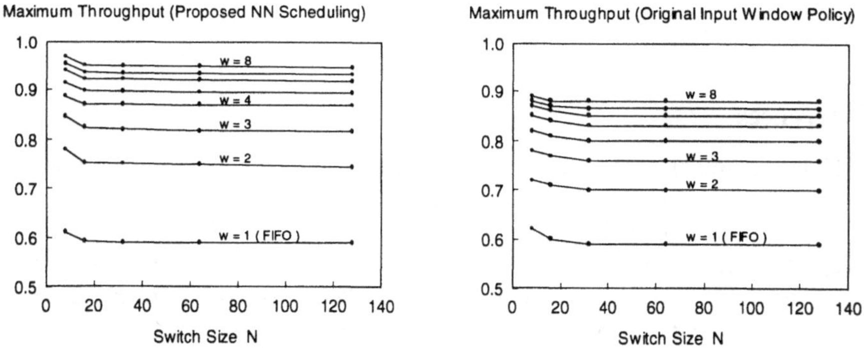

(a) (b)

Fig. 3.

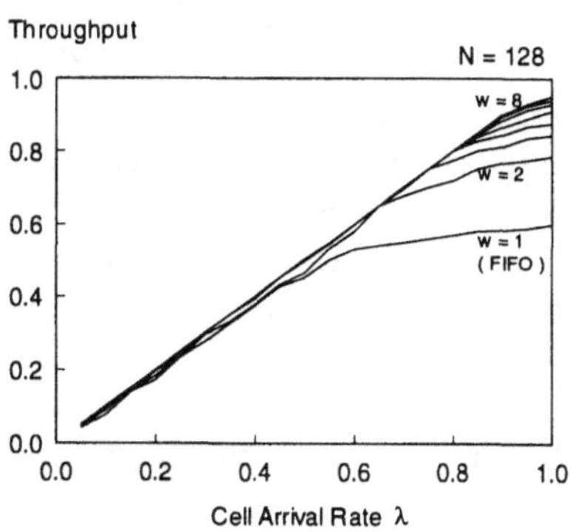

Fig. 4.

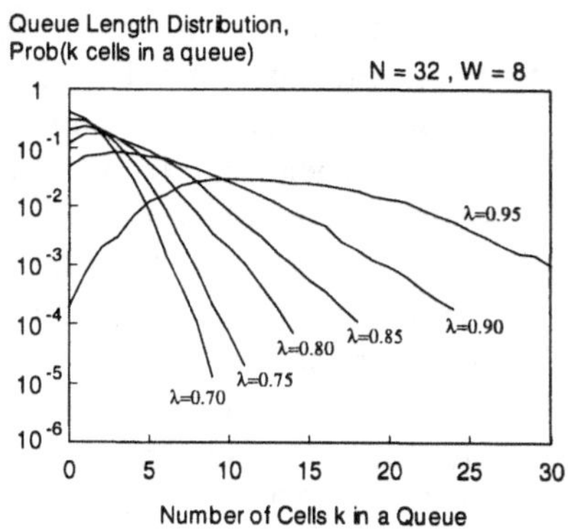

Fig. 5.

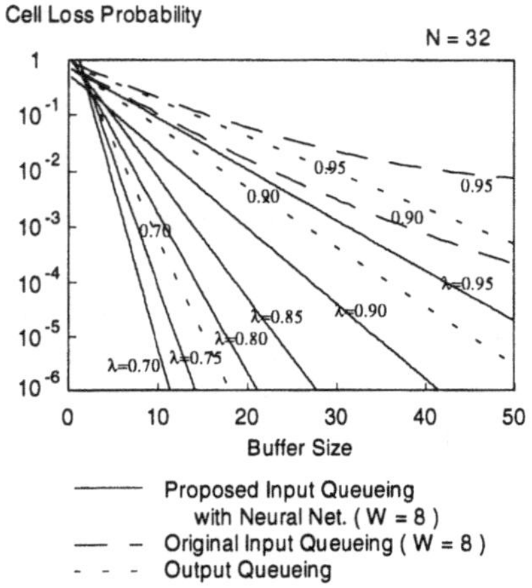

Fig. 6.

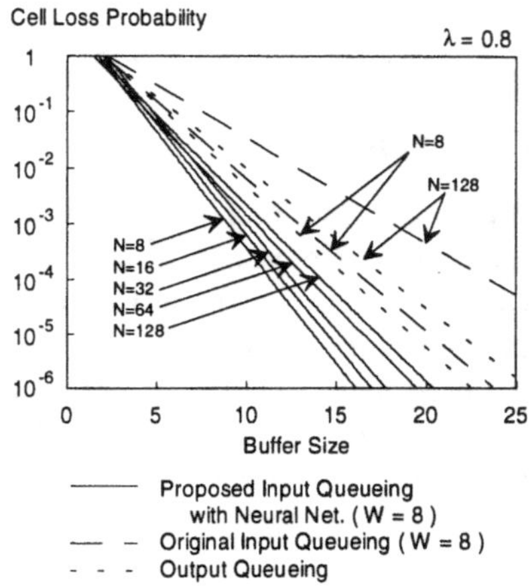

Fig. 7.

References

[1] H. Ahmadi, "A survey of modern high-performance switching techniques," *IEEE Journal on Selected Areas in Communications*, vol. 7, no. 7, pp. 1091-1103, Sept. 1989.

[2] M. J. Karol, M. G. Hluchyj, and S. P. Morgan, "Input versus output queueing on a space-division packet switch," *IEEE Trans. Communications*, vol. com-35, no. 12, pp. 1347-1356, Dec. 1987.

[3] M. G. Hluchyj and M. J. Karol, "Queueing in high-performance packet switching," *IEEE Journal on Selected Areas in Communications*, vol. 6, no. 9, pp. 1587-1597, Dec. 1988.

[4] T. T. Lee, "A modular architecture for very large packet switches," *IEEE Trans. Communications,* vol. 38, no. 7, pp. 1097-1106, July 1990.

[5] J. A. Freeman and D. M. Skapura, *Neural Networks, Algorithms, Applications, and Programming Techniques*, Addison-Wesley, 1991.

[6] J. J. Hopfield and D. W. Tank, "Neural computation of decisions in optimization problems," *Biological Cybernetics*, 52, pp. 141-152, 1985.

[7] Y. K. Park and V. Cherkassky, "Neural network controller for rearrangeable switching networks," *Int. Conf. Neural Networks (ICNN)*, San Francisco, vol. 3, pp. 1896-1901, Mar. 1993.

[8] A. Marrakchi and T. Troudet, "A neural network arbitrator for large crossbar packet-switches," *IEEE Trans. on Circuits and Systems*, vol. 36, no. 7, pp. 1039-1041, July 1989.

[9] T. Sziranyi and J. Csicsvari, "High-speed character recognition using a dual cellular neural network architecture," *IEEE Trans. on Circuits and Systems II*, vol. 40, no. 3, pp. 223-231, March 1993.

Using Genetic Algorithms to Design Constant Weight Codes

Francesc Comellas and Ramon Roca

Departament de Matemàtica Aplicada i Telemàtica
Universitat Politècnica de Catalunya
ETSE Telecomunicació, Campus Nord C-3, Gran Capitán s/n
E-08071 Barcelona, Catalonia, Spain
`comellas@mat.upc.es`

Abstract

Genetic algorithms have been used successfully for solving different combinatorial optimization problems. We give here a new efficient application for the search of constant weight codes. The algorithm presented is intended to generate codes with the maximum number of codewords for a given length, constant weight and minimum Hamming distance. A comparison with the simulated annealing technique is also discussed. On the other hand, the algorithm introduces a new scaling for the cost function that proved to be more effective than other scalings used in the literature.

1 Introduction

The design of good codes is of fundamental importance in a communication system. Information to be transmitted is represented via a *source code* as the ASCII code, Huffman code, etc. This source-encoded information is then sent over a channel, such as a telephone line, optical fiber, microwave link, etc. To have a reliable transmission the data are encoded again by using an *error-correcting code* that enables the detection and the correction of possible errors introduced during the transmission of the message (see Figure 1). Several modes of efficient codification are known: Hamming codes, Reed-Solomon codes, convolution codes, etc. The error-correcting code that

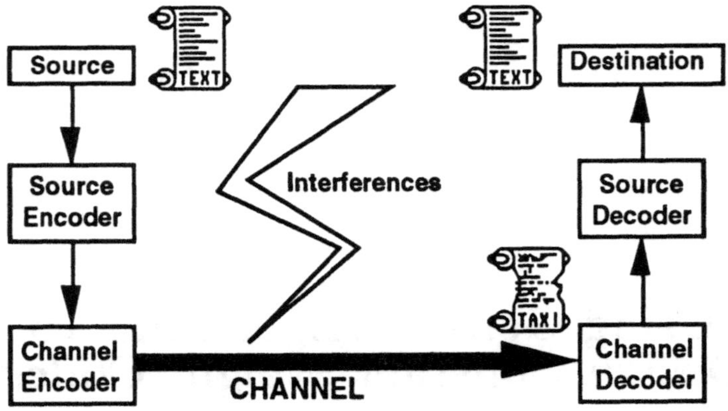

Figure 1: The coding process.

we consider in this paper is a set of binary n-tuples $C = \{x_1, x_2, \ldots, x_n\}$. Each member $w \in C$ is called a codeword. The Hamming distance $d_H(v, w)$ between two codewords is the number of bit positions where they differ. The minimum distance of C is $d_{\min} = \min_{\substack{v,w \in C \\ v \neq w}} d_H(v, w)$. When the error-correcting code is used in a communication system, up to $d_{\min}/2$ transmission errors can be detected and corrected. Therefore a large minimum distance $d_{\min}$ is a desirable property of an error-correcting code.

The Hamming weight of a codeword $w \in C$ is the number of ones in w. We consider only *constant-weight codes* where all codewords have the same Hamming weight.

The optimization problem studied consists of finding an error-correcting code of size as near as possible to the theoretical maximum size $A(n, d, w)$ for a fixed length n, minimum distance d (or more) and constant weight w.

Codes with size $A(n, d, w)$ are extremely difficult to find, see [1, 4]. So it is interesting to give simple methods that explicitly construct codes with size as high as possible and in this way to obtain a bound for $A(n, d, w)$. One of the methods that has proved useful is simulated annealing. El Gamal *et al.* in [2] describe the application of simulated annealing to several codification problems and in particular to the design of constant weight codes. In this paper we present a new method based on a genetic algorithm. The work of El Gamal *et al.* is used as a reference to compare both techniques. In Section 2 we describe the algorithm and also a new scaling procedure for the cost function that was developed for it. The performance of this scaling is compared with other known scalings. In Section 3 we give the results obtained and a comparison is made with the results coming from simulated annealing. Finally, in Section 4 the conclusions are presented.

2 The Genetic Algorithm

The genetic algorithm considered is based on the Simple Genetic Algorithm from the book of Goldberg [3]. Each generation has the same constant number of individuals and three genetic operators are used: reproduction, crossover and mutation.

The main aspects to decide are the representation of the solutions, the cost function and the crossover and mutation operators.

We are interested in the design of constant weight binary codes of size S, length L, weight w and minimum distance $d_{\min}$ or more. We code each possible solution by a list of $M = SL$ elements. Each position in the list has values 0 or 1. The list may be seen like the concatenation of all codewords from the corresponding code.

Cost function

The cost function used in the algorithm is:

$$f = K - \sum_{\substack{x,y \in C \\ x \neq y}} \frac{1}{g^2+1} \qquad g = \begin{cases} d_H(x,y) & \text{if } d_H(x,y) < d_{\min} \\ d_{\min} & \text{if } d_H(x,y) \geq d_{\min} \end{cases}$$

where K must prevent the cost function from taking negative values.

This cost function has a clear advantage over the cost function $f = \sum_{\substack{x,y \in C \\ x \neq y}} [d_H(x,y)]^{-2}$ used by El Gamal *et al.* [2]. Namely, in the El Gamal's function a large distance within two given codewords may compensate a short distance between two others and the total cost function be the same than in a code where all words are at equal distance. As a consequence, this function can not discriminate correctly between codes.

Exponential scaling

The study of standard scaling mechanisms used in the selection process lead us to the introduction of a new scaling, the *exponential scaling*:

$$s = e^{\frac{C(f-avg)}{max-avg}}$$

where f is the cost function and C a constant that enables the control of the probability of selection of an individual with maximum fitness in relation to an individual with average fitness. Usual values for C are between 2 and 3.

We have compared this new scaling mechanism with the linear, and rank scalings [3].

Figure 1 shows the maximum fitness at each generation (average of 50 runs) for the three scalings. The problem was the search of a code of size 26, length 23, minimum distance 10 and constant weight 9. A population of 300 individuals was considered. Other parameters were set as follows: $p_{cross} = 0.9$, $p_{mut} = 0.6$, $K = 20$ and $C = 2$. Similar performances were found for related problems.

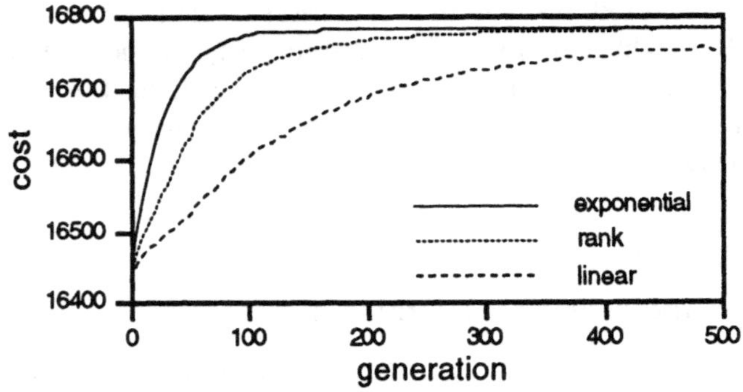

Figure 2: Exponential, rank and linear scalings compared.

The exponential scaling facilitates the control on the evolution of the algorithm, may deal with negative values of the cost function and is easy to implement. This new scaling might be also considered in other applications of genetic algorithms.

Crossover and mutation

The crossover and mutation operators were designed to deal with the characteristics of the problem. One important point is that all codewords must have the same constant weight and therefore, the genetic operators must preserve it.

The crossover on pairs of solutions $x^{par} = x_1x_2 \ldots x_M$ and $y^{par} = y_1y_2 \ldots y_M$, with $x_i, y_i \in \{0,1\}$, is done by randomly choosing a cutting point $r < M$, $r \bmod L = 0$, and creating the child lists as follows: $x^{chld} = x_1x_2 \ldots x_ry_{r+1} \ldots y_M$ and $y^{chld} = y_1y_2 \ldots y_rx_{r+1} \ldots x_M$. The crossover operator keeps full codewords from both codes but mixes the codes. The weights of the codewords are not modified. The fact that this crossover does not create new codewords may be seen as a drawback for the algorithm. To ensure a more extensive exploration of the state space, a special mutation operator was designed. Mutation is done by complementing the value of two different list elements, chosen at random, from positions corresponding to the same codeword. This operator preserves also the weights of the codewords.

3 Results

For our research we implemented a small program in C (less than 500 lines) based on the Pascal code presented in the book of Goldberg [3]. All experiments were performed on Sun SPARC2 and HP Apollo 730 workstations. After initial trials, the population size was set to 300, the crossover rate to 0.9 and the mutation rate to 0.6.

With these parameters we found the results shown in Table 1. Each code

Code	Simulated Annealing	Genetic Algorithm
A(21,10,9)	22 words	22 words
A(22,10,9)	23 words	**25** words
A(23,10,7)	18 words	*17* words
A(23,10,8)	28 words	*24* words
A(23,10,9)	24 words	**26** words
A(23,10,10)	39 words	39 words
A(23,10,11)	39 words	**40** words
A(24,10,8)	33 words	*30* words
A(24,10,9)	24 words	**27** words
A(24,10,11)	57 words	**58** words

Table 1: Codes found using Simulated Annealing and a Genetic Algorithm.

was obtained in the following way: Once fixed the code size, word length, minimum distance and constant weight the program was let running until the code was found (usually in less than 300 generations) or generation 1000 was attained.

As an example, Table 2 shows a code with 26 words, length 23, minimum distance 10 and weight 9.

The genetic algorithm found an equivalent result or a better code than the simulated annealing technique in most of the cases we studied. It was not possible to find three of the codes from El Gamal *et al.* paper. They have reported that these codes were found with "more computational effort" in order to improve some new results from Conway and Sloane, see [2]. The use of a more elaborated genetic algorithm would permit to obtain these codes and other codes that at present may be only constructed by special techniques as those cited in [1].

1.	11111101100000000000100	14.	00010001101000001101101
2.	01010001110000101010010	15.	00101011010110000100100
3.	00101110101001010001000	16.	11100001000100001101010
4.	10111000010000011100001	17.	00100000100011010110011
5.	10000000110101001110100	18.	11000110101000001010001
6.	00000100000110001011111	19.	00110110010101000000011
7.	00010100111010010010100	20.	00100101101100100110000
8.	01000001111101010000001	21.	10110000000100010011011
9.	00001100001000110100111	22.	01110010000110010011000
10.	00001010100110111000001	23.	01100101010010100001001
11.	10000110110000100001110	24.	00011001000101011000110
12.	00001000011011100111000	25.	10011011000011000001001
13.	01010110000011001100100	26.	11100000000100111000100

Table 2: A constant weight code A(23,10,9).

4 Conclusions

The results show that the genetic algorithm presented in this work perform well, and that in general, with less computational effort than in simulated annealing it is possible to obtain the desired codes. The main advantage of the method is its simplicity. On the other hand, the method is robust and may be easily implemented on a parallel computer.

Acknowledgement

This work was supported by the *CICYT*, Spain, under grant TIC-92-1228-E.

References

[1] A.E. Brower, J.B. Shearer, N.J.A. Sloane and W.D. Smith. "A new table of constant weight codes", *IEEE Trans. Inform. Theory*, vol. 36, pp. 134–138, 1990.

[2] A.A.El Gamal, L.A. Hemachandra, I. Shperling and V.K. Wei. "Using simulated annealing to design good codes", *IEEE Trans. Inform. Theory*, vol. 33, pp. 116–123, 1987.

[3] D. E. Goldberg. Genetic Algorithms in Search, Optimization, and Machine Learning, Addison-Wesley, 1989.

[4] R.L. Graham and N.J.A. Sloane. " Lower bounds for constant weight codes", *IEEE Trans. Inform. Theory*, vol. 26, pp. 37–43, 1980.

Bootstrap Methods in Neural Network Time Series Prediction

Jerome T. Connor

Bell Communications Research
Morristown, N.J. 07960 USA
connor@faline.bellcore.com

Abstract

Bootstrapping is a common technique for estimating quantities of a statistical model that are mathematically intractable. In this paper, we take a close look at the application of bootstrap techniques to the estimation of time series prediction error by neural networks. Bootstrap methods are shown to be useful in both prediction and the assessment of confidence intervals and prediction bias. In particular, we present an example involving access line data where bootstrap techniques are demonstrated.

One of the big problems in statistics is that a function, $h(\theta, \hat{\theta})$, of a true parameter, θ, and the estimate of the parameter, $\hat{\theta}$, from the sampled data, $x_1, \ldots, x_N$, can not be known exactly because θ is not known. Bootstrapping techniques are a method of estimating $h(\theta, \hat{\theta})$. The idea behind bootstrap methods is that $h(\theta, \hat{\theta})$ is likely to be similar to the function $h(\hat{\theta}, \tilde{\theta})$, where $\tilde{\theta}$ is is the estimated parameter obtained from data sampled from the observed data. How the resampling is accomplished will be described in section 2. The power of bootstrapping, as pointed out by Efron [1], is the ability to approximate $h(\hat{\theta}, \tilde{\theta})$ by simulations in situations where both $h(\theta, \hat{\theta})$ and $h(\hat{\theta}, \tilde{\theta})$ were both mathematically intractable. For an overview on bootstrap techniques see for example, Hall [5] and Efron [2].

There is a large body of bootstrap theory that is relevant to neural network time series prediction. For example, bootstrap techniques have been applied to regression problems by Freedman [3] and Wu [11]. Bootstrap techniques have also been applied to the time series domain by Freedman and Peters [4], Thombs and Schucany [10], and Stoffer and Wall [9]. As mentioned earlier, bootstrap methods are desirable when the quantities to be estimated are mathematically intractable. Consequently, some of the newer computationally intensive statistical techniques have attracted bootstrap methods such as nonparametric techniques, Hall [6], projection pursuit, Efron [2], and the field of neural networks Pass [8]. This paper will examine the role of bootstrapping for neural network time series predictors.

This paper will define a neural network one step ahead predictor for time series data in section 1. It will then be shown that the common way of iterating a one step ahead predictor to obtain predictions further into the future has drawbacks. In section 2 a new way of iterating a one step ahead predictor based on bootstrap techniques is defined. Confidence intervals and prediction bias are estimated by way of bootstrap techniques for the new predictor in section 3. Finally in section 4, these new methods are demonstrated on access line gain prediction, a problem that is relevant to the telecommunications industry.

1 Neural Network Time Series Model

Time series prediction in this paper is done via the nonlinear autoregressive, NAR(p), time series model

$$x_t = f(x_{t-1}, \ldots, x_{t-p}) + e_t \tag{1}$$

where f a nonlinear function, x_t is the observed time series, and e_t is a random process that can not be directly observed. The nonlinear function, f, will be modeled by a neural network to give the one step ahead predictor, $\hat{x}_{t|t-1,\ldots,t-p}$,

$$\hat{x}_{t|t-1,\ldots,t-p} = g(x_{t-1}, \ldots, x_{t-p}) = \sum_{i=1}^{H} W_i f(\sum_{j=1}^{p} w_{i,j} x_{t-j} + w_{i,0}) \tag{2}$$

where W_j and $w_{i,j}$ are parameters to be estimated from the data. For the rest of the paper, the parameters of the model will be referred to by the vector

$$\lambda = \{W_1, \ldots, W_H, w_{1,0}, \ldots, w_{1,p}, w_{2,0}, \ldots, w_{H,p}\} \tag{3}$$

which is determined from the data by

$$\hat{\lambda} = \underset{\lambda}{argmin} \sum_{t=p+1}^{N} (x_t - g(x_{t-1}, \ldots, x_{t-p}, \lambda))^2. \tag{4}$$

The predictor (2) can be used directly to make one step predictions. When predicting further into the future, one often uses future predictions as inputs into the neural network predictor in order to make an r step ahead predictor,

$$\hat{x}_{N+k} = \begin{array}{ll} g(\hat{x}_{N+k-1}, \ldots, \hat{x}_{N+1}, x_N, \ldots, x_{N+k-p}, \hat{\lambda}) & k \leq p \\ g(\hat{x}_{N+k-1}, \ldots, \hat{x}_{N+k-p}, \hat{\lambda}) & k > p \end{array} \tag{5}$$

This method has the advantage of only needing a single one step ahead predictor to determine all future predictors. If the predictor correctly models the true model, this predictor can yield good results, see [7]. However (5) suffers the drawback of making no allowances for future observations differing from predictions by a random elements, one of the assumptions of our model (1).

2 Bootstrap Time Series Prediction

In order to make realistic r step ahead predictions from (2), some estimate of future noise in trajectories must be obtained. To do this, we collect the set of estimated residuals, $e_i^{(0)}$, from the original model,

$$e_i^{(0)} = x_i - g(x_{i-1}, \ldots, x_{i-p}, \lambda^{(0)}) \tag{6}$$

where $\lambda^{(0)}$ is obtained from the original data, $x_1, x_2, \ldots, x_N$, by way of (4). The residuals calculated in (6) give rise to the empirical probability density, $P^{(0)}(e) = \frac{1}{N}\sum_{i=1}^{N} \delta(e - e_i^{(0)})$. One then synthesizes M possible future time series by randomly drawing new residuals, $e_k^{(j)}$ for $j = 1$ to M and $k = 1$ to r, and constructs a new time series via

$$\bar{x}_{N+k}^{(i)} = \begin{array}{ll} g(\bar{x}_{N+k-1}, \ldots, \bar{x}_{N+1}, x_N, \ldots, x_{N+k-p}, \lambda^{(0)}) + e_{N+k}^{(i)} & k \leq p \\ g(\bar{x}_{N+k-1}, \ldots, \bar{x}_{N+k-p}, \lambda^{(0)}) + e_{N+k}^{(i)} & k > p. \end{array} \tag{7}$$

A future prediction can then be made by

$$\hat{x}_{N+r} = median(\bar{x}_{N+r}^{(1)}, \ldots, \bar{x}_{N+r}^{(M)}). \tag{8}$$

When the residuals are not as the modeler would have expected, perhaps non-Gaussian or skewed, the simulation procedure described in (7) and (good) will outperform analytic methods based on the model (2) and an incorrect preconceived residual probability distribution. This is due to the use of the empirical distribution to describe the residuals.

3 Bootstrap Estimates of Confidence Intervals

A bootstrap estimate of the confidence intervals of a predictor (8) will now be obtained. First, a set of synthetic time series must be obtained via

$$\tilde{x}^{(i)}_{p+k} = \begin{array}{ll} g(\tilde{x}_{p+k-1}, \ldots, \tilde{x}_{p+1}, x_p, \ldots, x_k, \lambda^{(0)}) + e^{(i)}_{p+k} & k \leq p \\ g(\tilde{x}_{p+k-1}, \ldots, \tilde{x}_k, \lambda^{(0)}) + e^{(i)}_{p+k} & k > p \end{array} \tag{9}$$

where the residuals, $e^{(i)}_{p+k}$, are drawn from the empirical distribution described previously in section 2. New models, $g^{(j)}$, are constructed by estimating parameters, $\lambda^{(j)}$, from the synthesized time series,

$$\hat{\lambda}^{(j)} = \begin{array}{c} argmin \\ \lambda \end{array} \sum_{t=p+1}^{N} (\tilde{x}^{(j)}_t - g(\tilde{x}^{(j)}_{t-1}, \ldots, \tilde{x}^{(j)}_{t-p}), \lambda)^2. \tag{10}$$

The bootstrap estimate of the confidence intervals of the next step predictor is given by predicting the original time series, $x_1, \ldots, x_N$, with the models obtained from synthetic data, $g^{(j)}$, for the next step predictor

$$\hat{x}^{(j)}_i = g^{(j)}(x_{i-1}, \ldots, x_{i-p}, \lambda^{(j)}). \tag{11}$$

The confidence intervals are then obtained by estimating

$$P(\hat{x}_i - \hat{x}^{(j)}_i > \epsilon_H) = .95 \tag{12}$$

$$P(\hat{x}_i - \hat{x}^{(j)}_i < \epsilon_L) = .05. \tag{13}$$

The probability equations (12) and (13) are solved empiricly for ϵ_H and ϵ_L by examining a large number of bootstrap estimators.

For the r step ahead predictor, equation (11), must be altered to the form

$$\hat{x}^*_{N+r} = median(\hat{x}^{*(1)}_{N+r}, \ldots, \hat{x}^{*(M)}_{N+r}). \tag{14}$$

where the r step ahead prediction of the bootstraped model is given by

$$\hat{x}^{*(i)}_{N+r} = median(x^{*(i,1)}_{N+r}, \ldots, x^{*(i,M)}_{N+r}) \tag{15}$$

and

$$x^{*(i,j)}_{N+k} = \begin{array}{ll} g(x^{*(i,j)}_{N+k-1}, \ldots, x^{*(i,j)}_{N+1}, x_N, \ldots, x_{N+k-p}, \lambda^{(i)}) + e^{(i,j)}_{N+k} & k \leq p \\ g(x^{*(i,j)}_{N+k-1}, \ldots, x^{*(i,j)}_{N+k-p}, \lambda^{(i)}) + e^{(i,j)}_{N+k} & k > p \end{array} \tag{16}$$

and the residuals are drawn from the empirical distribution, $P(e^{(i,j)}) = \frac{1}{N}\sum_{k=1}^{N}\delta(x_k - \hat{x}_k^{(i)})$. Note that each of the M bootstraped predictors must be iterated individually to perform r step ahead prediction. The confidence intervals are then obtained by solving (12) and (13) as in the case of the one step ahead predictor.

In addition to defining an estimator and confidence intervals, bootstrap methods can also estimate the bias of a prediction, $\mu = E(x_{N+r} - \hat{x}_{N+r})$. This is obtained by using the assumption of bootstrap methods

$$E(x_{N+r} - \hat{x}_{N+r}) \approx E(\hat{x}_{N+r} - \hat{x}^{*}_{N+r}). \tag{17}$$

Combining (8), (14), and (17) yields the bootstrap estimate of the bias,

$$\hat{\mu} = median(\bar{x}^{(1)}_{N+r}, \ldots, \bar{x}^{(M)}_{N+r}) - median(x^{*(i,1)}_{N+r}, \ldots, x^{*(i,M)}_{N+r}). \tag{18}$$

It is then possible to define the unbiased predictor by

$$\hat{x}^{unbiased}_{N+r} = \hat{x}_{N+r} + \hat{\mu}. \tag{19}$$

4 Example on Access Line Data

The time series, shown in Figure 1, will be used to demonstrate the bootstrapping methods described earlier. This time series represents a transformation of the demand for access lines of a regional holding company. All scale and date information of the time series has been withheld for reasons of confidentiality.

In Figure 2 the predictor of the time series, which is given by the equations (7), (8), and (19), is a solid line plotted alongside the actual time series points. The 90% confidence intervals are also plotted as lines on either side of the prediction line, this will give the modeler some idea of the reliability of the model. The part of Figure 2 that is just one line corresponds to before the prediction. It is readily observed that the distance between the confidence intervals grows with the length of the prediction.

5 Conclusion

The bootstrap represents an intuitively pleasing computational algorithm to infer a quantity of a statistical model. This paper has shown how the bootstrap can be applied to neural networks to improve the prediction and assess the confidence intervals and bias of a predictor. Bootstrap methods have been demonstrated to perform well on time series that are relevant to the telecommunications industry.

Acknowledgements

The author would like to thank R. Douglas Martin for talks that led to this paper.

References

[1] Efron, B. " Bootstrap Methods: Another Look at the Jackknife", *Annals of Statistics*, 1979.

[2] Efron, B. and Tibshirani, R., "Bootstrap Methods for Standard Error", Confidence Intervals, and Other Measures of Statistical Accuracy, *Statistical Science*, Vol. 1, No. 1, pp. 54-77, 1986.

[3] Freedman, D. A. "Bootstrapping Regression Models", *Annals of Statistics*, Vol. 9, No. 6, pp 1218-28, 1981.

[4] Freedman, D. A. and Peters, Stephen C. "Bootstrapping a Regression Equation: Some Empirical Results", *J. of the Amer. Stat. As.*, Vol. 79, No. 385, March 1984, pp. 97-106.

[5] Hall, P. *The Bootstrap and Edgeworth Expansion*, Springer-Verlag, New York, 1992.

[6] Hall, P. "On Bootstrap Confidence Intervals in Nonparametric Regression", *Annals of Statistics*, Vol. 20, No. 2, pp. 695-711, 1992.

[7] Lapedes, A. and Farber, R., "Nonlinear Signal Processing Using Neural Networks: Prediction and Modeling", Technical Report, LA-UR87-2662, Los Alamos National Laboratory, Los Alamos, New Mexico, 1987.

[8] Pass, Gerhard. "Assessing and Improving Neural Network Predictions by the Bootstrap Algorithm." In *Advances in Neural Information Processing Systems 5.* Morgan Kaufmann, Hanson, J., Cowan, J., and Giles, C. eds. San Mateo, California. 1993.

[9] Stoffer, D. S. and Wall, K. D. "Bootstrapping State-Space Models: Gaussian Maximum Likelihood Estimation and the Kalman Filter", *J. of the Amer. Stat. As.*, Vol. 86, No. 416, pp. 1024-33, 1991.

[10] Thombs, L.A. and Schucany, W.R. "Bootstrap Prediction Intervals for Autoregression",*J. of the Amer. Stat. As.*, Vol. 85, no. 410, pp. 486-92, 1990.

[11] Wu, C.F.J. "Jackknife, Bootstrap and other Resampling Methods in Regression Analysis", *Annals of Statistics*, Vol. 14, No. 4, pp. 1261-95, 1986.

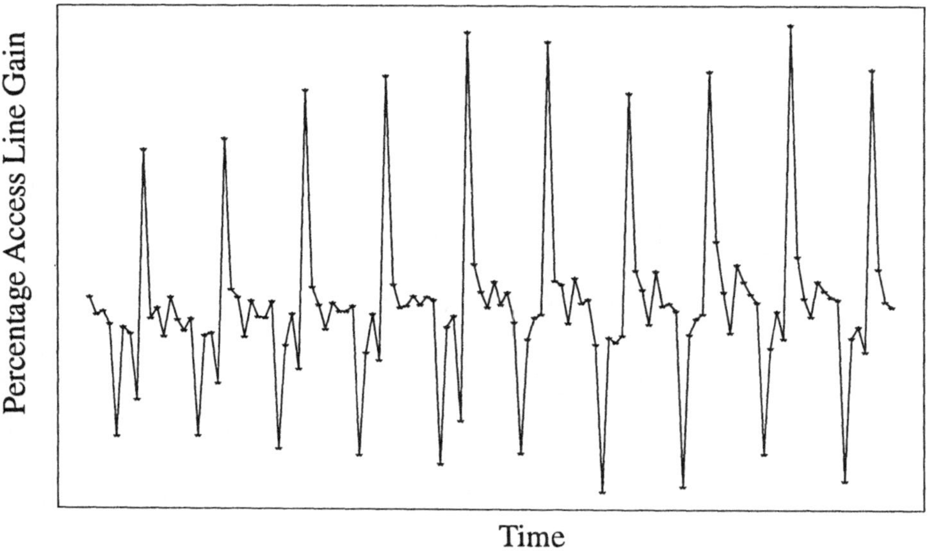

Figure 1: The access line gain vs time.

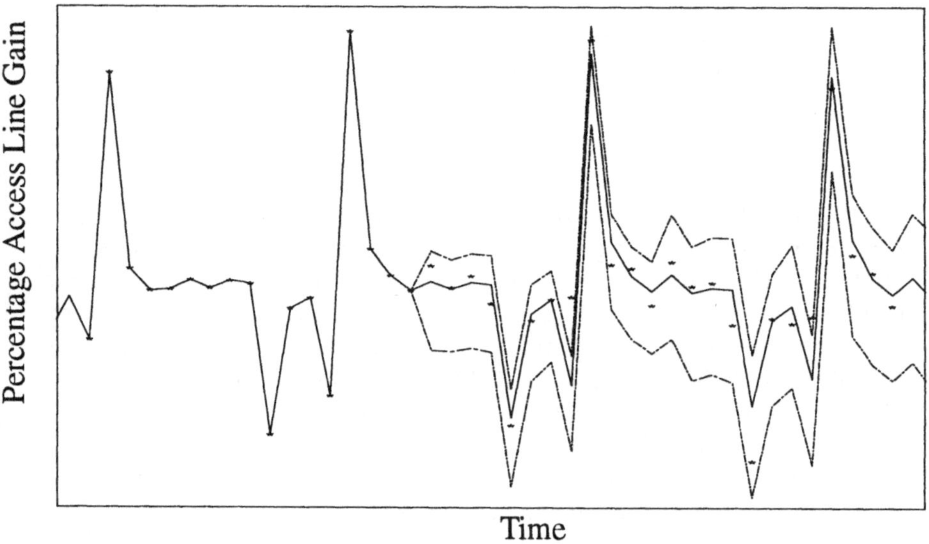

Figure 2: A close up view of the access line gain, neural network predictions, and confidence bars vs time.

16kb/s Adaptive Differential Pulse Code Modulation of Speech

S. Haykin *and* **L. Li**
Communications Research Laboratory, McMaster University
Hamilton, Ontario, Canada L8S 4K1
haykin@mcmaster.ca liang@nimios.mcmaster.eng.ca

Abstract

In this paper we describe a novel nonlinear adaptive differential pulse code modulation (ADPCM) system capable of coding speech at 16 kb/s. The nonlinear predictor involves the use of two subsections: (1) a nonlinear structure consisting of a cascade of two real-time recurrent neural networks, and (2) a linear predictor.

1 Introduction

Predictive coding has established itself as a useful technique for the digital representation of speech signals at a rate lower than the standard pulse code modulation(PCM). To accomplish this reduction in bit rate, a combination of adaptive predictive encoding and adaptive quantization is used. The resulting technique is commonly referred to as adaptive differential pulse code modulation(ADPCM)[1]. Indeed, 32 kb/s ADPCM is presently recognized, alongside 64 kb/s PCM, as an international standard by CCITT.

The conventional form of ADPCM relies on the use of a linear predictor, yet it is widely recognized that the biological mechanism responsible for the generation of speech signals is in fact nonlinear. This points to a shortcoming of the conventional form of ADPCM. In this paper we address this issue by investigating the use of a neural network as the basis for the design of a nonlinear predictor. However, the choice of a suitable neural network architecture for such an application is constrained by the fact that the network has to perform its learning in an **on-line** fashion as the speech signal is being continuously processed. In this paper, we describe a novel neural network that accomplishes this requirement.

2 Real-time Nonlinear Adaptive Predictor

The predictor of a conventional ADPCM system uses the least-mean-square (LMS) algorithm for the adaptation of a linear, infinite (duration) impulse response (IIR) filter. The stability and mistracking recovery of the encoder/decoder are given particular attention. In the linear 32 kb/s ADPCM system, the same linear adaptive predictor is used in both the transmitter and the receiver. The adaptive filter used therein has a transfer function consisting of two poles and six zeros; a novel weight-update algorithm is employed to minimize mistracking behavior of this linear predictor[2].

A great deal of research effort has been expended in the design of such a structure with the aim of matching it to the characteristics of speech signals. It is therefore natural that we build on this valuable experience. To do so, we have developed a neural network-based nonlinear predictor that consists of two subsections:

- A nonlinear subsection made up of two identical modules connected in cascade, and with each module consisting of a real-time recurrent neural network[3].
- A linear subsection that is of the same design as the linear predictor in the standard ADPCM[2], but whose weight update algorithm is actuated by outputs derived from the two modules.

The modeling approach used for the predictor design is thus locally linear but globally nonlinear[4].

Figure 1(a) presents a block diagram of the nonlinear predictor described here. Both modules are designed to have exactly the same set of synaptic weights. The two outputs of Module 2 are applied to Module 1, and also (after a delay of one time unit) they are fed back to Module 2. Each module consists of two neurons, with an input layer made up of 3 external(source) nodes and 2 feedback nodes (one corresponding to each neuronal output), as shown in Figure 1(b). The node assigned a fixed input equal to 1 is included for the purpose of applying a bias to each of the neurons. The output vector $\mathbf{y}_o(n)$, consisting of output $y_1(n-1)$ of Module 1 and output $y_1(n-2)$ of Module 2, is applied to a linear adaptive predictor to produce an estimate of the present value of the input speech signal. The algorithm for on-line training and operation of the nonlinear subsection is summarized in Table 1[5]. The design of the linear subsection is similar to that of the linear predictor in the 32 kb/s ADPCM.

Table 1: *Algorithm for the on-Line Training and Nonlinear Prediction*

- **Initialization,** initialize the algorithm by setting

$$\mathbf{r}_{fb}(n-2) = \mathbf{y}, \qquad \mathbf{W} = \mathbf{W}_0$$

where $\mathbf{y}$ is the output vector and $\mathbf{W}_0$ is the weight matrix of the epochwise-trained module with n_e samples of the input signal[3].

- **Training Part**

For each instant of time, input $r(n)$, $n \geq n_e + 1$, and compute the predicted output vectors of Modules 1 and 2:

$$\begin{aligned}\hat{\mathbf{y}}(n-2) &= [\hat{y}_1(n-2), \hat{y}_2(n-2)]^T \\ \hat{\mathbf{y}}(n-1) &= [\hat{y}_1(n-1), \hat{y}_2(n-1)]^T\end{aligned}$$

- Update the weight matrix $\mathbf{W}$ of Modules 1 and 2 by minimizing the cost function

$$\varepsilon(n) = \sum_{i=1}^{2} \lambda^{(i-1)}(d(n-i) - \hat{y}_1(n-i))^2$$

where $1 < \lambda < 1$. In the real-time recurrent neural network training algorithm[3], gradient descent is used to minimize the cost function $\varepsilon(n)$. Let the $\Delta\mathbf{W}$ denote the computed change in the weight matrix according to this algorithm. The updated weight matrix is

$$\mathbf{W} \leftarrow \mathbf{W} + \Delta\mathbf{W}$$

- **Operation Part**

Calculate the estimated output vector of Modules 1 and 2:

$$\mathbf{y}(n-2) = [y_1(n-2), y_2(n-2)]^T$$

$$\mathbf{y}(n-1) = [y_1(n-1), y_2(n-1)]^T$$

- **Linear Prediction Part**

Overall predicted output:

$$\begin{aligned}\hat{r}(n+1) &= \mathbf{a}^T\mathbf{y}_o(n) + \mathbf{b}^T\hat{\Delta}(n) \\ \mathbf{y}_o(n) &= [y_1(n-1), y_1(n-2)]^T\end{aligned}$$

where the weight vector $\mathbf{a}$ is for modeling poles and the weight vector $\mathbf{b}$ is for modeling zeros in the input signal, and $\hat{\Delta}(n)$ is a vector of delayed transmitted signals. Using the linear adaptive algorithm described in [2], update the weight vectors $\mathbf{a}$ and $\mathbf{b}$.

- **Recursive Calculation,** set

$$n = n + 1, \qquad \mathbf{r}_{fb}(n-2) = \mathbf{y}((n-1)-2)$$

Return to the **training part** of Modules 1 and 2.

3 Nonlinear ADPCM system

In using a nonlinear adaptive predictor, we are in effect constituting a new ADPCM system. The neural network-based nonlinear predictor described in Section 2 is used to perform the prediction in the encoder and decoder. The quantizer is a two-level backward adaptive quantier. A block diagram illustrating the structure of the 16 kb/s nonlinear ADPCM system is shown on Figure 2.

The speech signal $s(n+1)$ is input to the nonlinear ADPCM encoder. The nonlinear predictor forms an estimate $\hat{s}(n+1)$ of the signal's present value based on the past reconstructed signals $r(n), r(n-1), \cdots$ at the nonlinear ADPCM encoder. The estimate $\hat{s}(n+1)$ of the signal is then subtracted from the signal $s(n+1)$ to form the difference signal $\delta(n+1)$, which is quantized, encoded, and transmitted to the decoder of the receiver as $q(n+1)$. At the same time, the delayed transmitted signal $q(n)$ is decoded at the transmitter. The reconstructed signal $r(n)$, is then used to predict the next sample of the input signal.

At the receiver, the transmitted signal $q(n+1)$ is decoded, and added to the estimate $\hat{s}(n+1)$ of the signal to form the reconstructed speech signal $\hat{\hat{s}}(n+1)$. The nonlinear predictor used in the decoder is identical to the one employed in the encoder.

4 Experimental Results

We have tested the nonlinear ADPCM system described herein with various speech signals. In this Section we present highlights of this experimental work based on a male speech: **when we record audio data...** . The recorded time series corresponding to this speech signal, sampled at 8 kHz, is made up of 10,000 points. Specifically, we may say the following:

1. The nonlinear prediction of the speech signal in the encoder part of the system is shown as the dashed curve in Figure 3(a); the actual speech signal is shown by the continuous curve. Figure 3(b) shows the power spectrum of the resulting prediction error, which is fairly

constant across a band of 4000 Hz. Moreover, the probability distribution of the prediction error (not shown) approximates a Gaussian distribution fairly closely. The conclusion to be drawn from these observations is that the (nonlinear) prediction error may be closely modeled as a white Gaussian noise process. This is testimony to the efficiency of the neural network-based predictor in extracting the information content of the speech signal almost fully.

2. Figure 4 presents a comparison of the reconstruction performance of the 16 kb/s nonlinear ADPCM system with that of the conventional 32kb/s linear ADPCM system. These results are shown in parts (a) and (b) of Figure 4, respectively; the continuous curves pertain to the actual speech signal, and the dashed curves pertain to the nonlinear and linear reconstructed versions as measured at the decoder part of the system. The results described in Figure 4 clearly show that the reconstructed signal using a nonlinear ADPCM provides a better approximation to the actual speech signal than the linear ADPCM, even though the linear system has the advantage of operating at twice the bit rate.

3. For a quantitative evaluation of reconstruction performance, we may use the following error measure expressed in decibels:

$$R_p = 10 * \log_{10}(\delta_r^2/\delta_s^2)$$

where δ_s^2 is the average-mean-square value of the speech signal at the transmitter input, and δ_r^2 is the corresponding value of the reconstruction error at the output of decoder. For 10,000 speech samples, δ_s^2 is about 0.6706. For 16kb/s nonlinear ADPCM, δ_r^2 is about $8.9 * 10^{-4}$, yielding $R_p = -28.9\text{dB}$. On the other hand, for the linear 32kb/s ADPCM, δ_r^2 is about $16.3 * 10^{-4}$, yielding $R_p = -26.1\text{dB}$. This provides a quantitative proof of the superior performance of the nonlinear 16 kb/s ADPCM over the linear 32 kb/s ADPCM.

4. Listening tests show that there is only a slight degradation in the quality of the reconstructed speech signal using the 16 kb/s nonlinear ADPCM system, and that the quality is slightly better than that of 32 kb/s linear ADPCM.

5. For processing the male speech signal, the computational time of 16 kb/s nonlinear ADPCM is just twice as long as that of 32 kb/s linear ADPCM.

5 Concluding Remarks

¿From the experimental results just presented, we conclude that the overall performance of the 16 kb/s nonlinear ADPCM system is slightly better than that of the 32 kb/s linear ADPCM, even though the linear system has the advantage of operating at twice the bit rate.

According to a rule of thumb reported in [1], a reduction in the bit rate of speech coding by a factor of two using conventional techniques is achieved at the cost of a ten-fold increase in hardware complexity. In contrast, the experimental results presented in this paper show that we can indeed build a 16 kb/s ADPCM system using a nonlinear predictor (based on neural network technology) with a performance comparable to the standard 32 kb/s ADPCM, and yet the increase in hardware complexity is minor compared to that of a 16 kb/s sub-band coding system.

References

[1] N.S. Jayant, "Coding Speech at low bit rates," *IEEE Spectrum,* August, 1986.

[2] N.Benvenuto, "The 32-kb/s ADPCM Coding Standard," *AT&T Technical Journal,* Vol 65,Issue 5. September/October, 1987.

[3] R.J.Williams and D.Zipser, "A Learning Algorithm for continually Running Fully Recurrent Neural Networks," *Neural Computation,* Vol 2, pp270-280, 1987.

[4] M.B.Priestley, "Non-linear and Non-stationary Time Series Analysis," ACADEMIC PRESS. 1988, pp.91-137.

[5] Liang Li and Simon Haykin, "A Cascaded Recurrent Neural Networks for Real-Time Nonlinear Adaptive Filtering," in Proceedings of the Int'l Conference on Neural Networks, San Francisco, 1993, pp857-862.

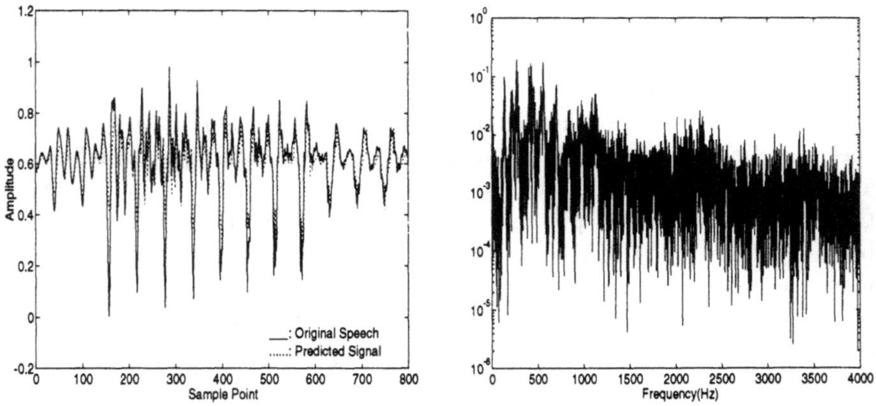

(a) Prediction of speech in the encoder of nonlinear predictor

(b) Power spectrum of the nonlinear prediction error

Figure 3:

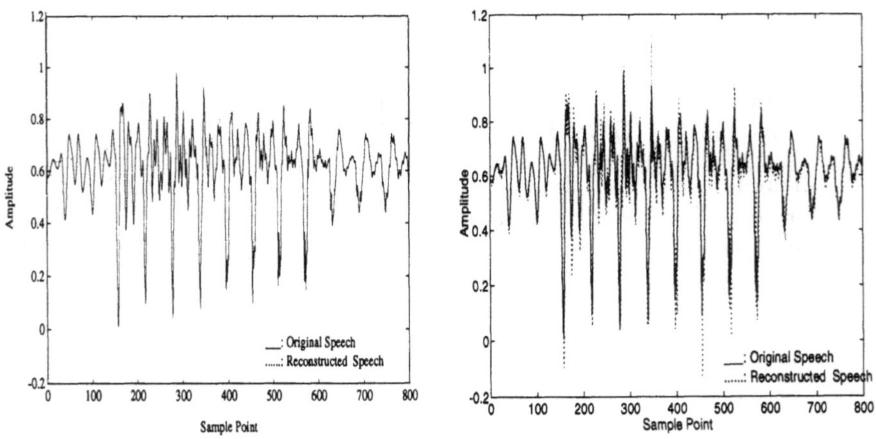

(a) Reconstruction of speech with 16 kb/s nonlinear ADPCM

(b) Reconstruction of speech with 32 kb/s linear ADPCM

Figure 4:

Adapting-Size Neural Network for Character Recognition on X-ray Films

Akira Hasegawa, Kenji Shibata†, Kazuyoshi Itoh†, Yoshiki Ichioka†, and Kiyonari Inamura††

ISIS Center, Department of Radiology,
Georgetown University Medical Center
3800 Reservoir Rd., NW , Washington, DC 20007, USA
†Department of Applied Physics, Osaka University, JAPAN
†† College of Biomedical Technology, Osaka University, JAPAN

Abstract

In this paper, we propose a novel learning method for category learning. The proposed method not only adjusts weights and biases but optimizes the number of hidden units. We applied this method to character recognition on X-ray films and achieved a recognition rate of 99.4% on a 1,200-element testing set.

1 Introduction

Neural networks have been recognized as powerful tools for signal processing. Neural networks obtain transfer functions by learning weights and biases. One of the popular learning algorithms is the error back propagation (EBP) algorithm [1]. It has been successfully applied to realistic problems [2, 3]. However, many researchers point out the following shortcomings; empirical determination of an optimal network architecture, slow learning process, and trapping at local minima. RCE (Restricted Coulomb Energy) method [4] is known for a powerful algorithm for category learning. The RCE adjusts the number of hidden units and their biases of three-layered feedforward networks. Advantages of RCE are the automatic determination of the number of hidden units, the capabilities of quick, and additional learning. However, it has a shortcoming that the number of hidden units is apt to increase excessively if training examples include random noise. This is because weights between the input and hidden units are selected from training examples and are never improved even though they're not optimal.

To solve the problems mentioned above, we propose a learning algorithm, HSII (Hyper-spherical Surface Interactive Interconnections) method for category learning. This algorithm automatically adjusts the number of hidden units and their biases and improves weights between the input and hidden units. The weights between the hidden and output units are fixed.

We applied our HSII to recognition of characters on medical X-ray films. The recognition of characters on X-ray films, such as ID numbers of patients, is dispensable technique for storing and sending medical X-ray films in hospitals. Fast and reliable method for the character recognition is desired.

2 Algorithm of the HSII

Let $\boldsymbol{p}^i$ be N-dimensional pattern vectors which can be categorized into C classes. $\boldsymbol{p}^i(c)$ represents vector $\boldsymbol{p}^i$ falling into cth class. A three-layered feedforward neural network consisting of N input units, H hidden units, and C output units, is utilized for the classification. The output patterns are trained in such a way that cth output unit is fired when $\boldsymbol{p}^i(c)$ is presented on the input layer. A training set, $\{(\boldsymbol{p}^i(c), \boldsymbol{t}^i(c)) \mid i = 1, 2, ..., m_t\}$, is used for the training. Every input vectors are normalized to satisfy

$$\boldsymbol{p}^i(c) \cdot \boldsymbol{p}^i(c) = L, \tag{1}$$

and $\boldsymbol{t}^i(c)$ is a C-dimensional target vector whose cth component, t_c^i, is 1 and the other components are 0.

When $\boldsymbol{p}^i(c)$ is represented to the neural network, the value received at the hth unit in the hidden layer, $O_h(\boldsymbol{p}^i)$, is given by

$$O_h(\boldsymbol{p}^i) = f_1[\boldsymbol{p}^i(c) \cdot \boldsymbol{w}_h - b_h], \tag{2}$$

where $\boldsymbol{w}_h = (w_{1,h}, w_{2,h}, ..., w_{n,h}, ..., w_{N,h})$ denotes a normalized weighting vector between the input units and hth hidden unit. The b_h is the bias of hth hidden unit. The transfer function on the hidden layer $f_1(\cdot)$ is,

$$f_1(x) = \begin{cases} k_1 x & (x > 0) \\ \dfrac{2}{1 + exp(-k_1 x)} - 1 & (x \leq 0). \end{cases} \tag{3}$$

The weighting vector associated with the connection between the hth hidden unit and the output units is $\boldsymbol{v}_h = (v_{h,1}, v_{h,2}, ..., v_{h,c}, ..., v_{h,C})$. The output value of the cth unit in the last layer, $R_c(\boldsymbol{p}^i)$, is given by

$$R_c(p^i) = f_2\left[\sum_{h=1}^{H} O_h(p^i) \cdot v_{h,c}\right], \tag{4}$$

where $f_2(\cdot)$ denotes the transfer function of the output units and is defined as

$$f_2(x) = \begin{cases} \dfrac{2}{1 + exp(-k_2 x)} - 1 & (x \leq 0) \\ 0 & (x > 0). \end{cases} \tag{5}$$

The output value, $R_c(\boldsymbol{p}^i)$ generated from the neural network from a given input vector may or may not identical to the desired output value $\boldsymbol{t}^i(c)$. The sum of the squared-error for all output units is:

$$E_i = \frac{1}{2} \sum_{c=1}^{C} \left[R_c(p^i) - t_c^i\right]^2, \tag{6}$$

which is used in EBP. In this method, we define the error signal of hth hidden unit δ_h as

$$\delta_h = f_1'(\boldsymbol{p}^i \cdot \boldsymbol{w}_h - b_h) \sum_{c=1}^{C} (R_c - t_c^i) \cdot v_{h,c}. \tag{7}$$

which is equivalent to the error propagation $\delta_h(\mathrm{EBP})$ derived from Eq. (6) for the EBP method.

$$\delta_h(EBP) = f_1'(\boldsymbol{p}^i \cdot \boldsymbol{w}_h - b_h) \sum_{c=1}^{C} (R_c - t_c^i) \cdot v_{h,c} \cdot f_2\left[\sum_{h=1}^{H} O_h(p^i) \cdot v_{h,c}\right]. \tag{8}$$

The difference between Eqs. (7) and (8) is the last component $f_2'(\cdot)$ which is always equal to 1 in the HSII.

During the HSII training, H, $\boldsymbol{w}_h$, and b_h are simultaneously updated. The weights $\boldsymbol{v}_h$ do not participate in the training processes, however, they must coincide to the fired output units. The HSII has two learning modes: (i) improving $\boldsymbol{w}_h$ and b_h by back-propagation of errors, and (ii) adding new hidden unit. When $\boldsymbol{p}^i(c)$ is presented, the selection of these learning modes depends on both the activity of cth output unit and the absolute values of the error signals δ_hs of hidden units connected to cth output unit by non-zero weights. When the cth output unit doesn't fire and $|\delta_h|$ is small, learning mode (ii) is active. Otherwise, learning mode (i) is taken place. This is described more specifically as the following:

Once an input vector $\boldsymbol{p}^i(c)$ is presented to the network, R_c and $|\delta_h|$ can be calculated as indicated in equations (4) and (7), respectively. The active training mode is determined to be (i) by

$$\begin{gathered} R_c\left[\boldsymbol{p}^i(c)\right] \neq 0, \\ |\delta_h| > \mu_0, \end{gathered} \tag{9}$$

where μ_0 denotes a small positive constant. When mode (i) is selected, the weights and biases are renewed by Eq. (10),

$$\begin{gathered} \overline{\boldsymbol{w}_h}(t+1) = \boldsymbol{w}_h(t) - \eta_w \cdot \boldsymbol{p}^i \cdot \delta_h, \\ b_h(t+1) = b_h(t) - \eta_b \cdot \delta_h, \end{gathered} \tag{10}$$

where η_w and η_b are learning rates of weight vector and bias, respectively. Improved weights are normalized by,

$$w_h(t+1) = \frac{\overline{w_h}(t+1) \times L}{\sum_{n=1}^{N} \overline{w_{n,h}}(t+1)} \tag{11}$$

When mode (ii) is selected, R_c doesn't fire and δ_hs are nearly equal to 0. In this case, the H+1th hidden unit is created. The newly generated parameters, w_{H+1}, b_{H+1}, and v_{H+1} are initialized as follows;

$$\begin{aligned} w_{H+1} &= p^i(c), \\ v_{H+1} &= t^i(c), \\ b_{H+1} &= \lambda. \end{aligned} \tag{12}$$

where λ is a constant. The creation of H+1th hidden unit can make cth output unit equal to t_c^i.

The initial network has no hidden units. When the first training example is presented to the initial network, the learning mode (ii) is selected. Above mentioned learning modes are iteratively executed on all training examples until the total sum of errors,

$$\varepsilon = \sum_{i=1}^{m_i} \sum_{c=1}^{C} \left| R_c(p^i) - t_c^i \right|, \tag{13}$$

is smaller than ε_0.

3 Character Recognition by the HSII

Fig. 1(a) shows an example of original X-ray images. Ten numerals, 0-9, and three alphabets, T, M, and S which mean the names of Japanese eras, on the first, third, and forth lines are to be recognized. On the second line names of patients are recorded, but they are concealed for guarding the privacy in Fig. 1(a).

(a) (b)

Fig. 1. (a) Example of original X-ray images and (b) preprocessed and extracted characters from original X-ray films

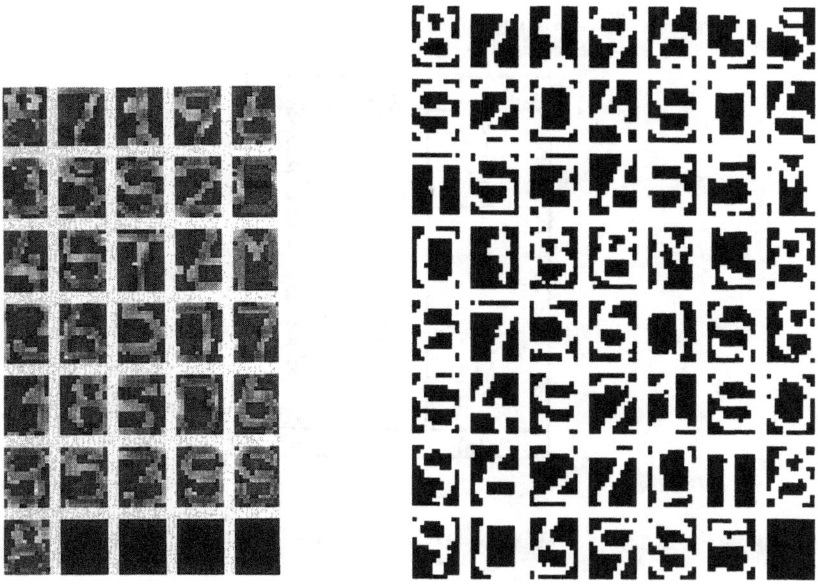

(a) (b)

Fig. 2. (a) The created weight images in the learning process of the HSII. The number of learning steps for the algorithm converged on a 2,250-elements training set was 710, and (b) the created weight images in the learning process of the RCE.

In Fig. 1(b), we show the characters extracted from the original image shown in Fig. 1(a). Every character consists of binarized 12×15 pixels which are +1 or -1. In preprocessing, the original images were rotated and binarized. To recognize these inputs, the neural networks have 180 (12×15) input units and 13 output units.

With a set of parameters obtained empirically; η_b=0.2, η_w=0.005, λ=0.0, k_1=k_2=0.5, and ε_0=0.2, we trained the network by using a 2,250-element training set extracted from 90 sheets of X-ray images. The number of learning iterations for the convergence was 710 steps, and 31 hidden units were generated during the learning process. For the comparison, we also trained a network with the same size by the EBP algorithm. The EBP could not converge even after 12,000 steps. This comparison shows the faster convergence of the HSII.

The created weight images in the learning process of the HSII are shown in Fig. 2 (a). To compare with the RCE, we trained the RCE network by the same training set. In the RCE, 55 hidden units were generated and the weight images are shown in Fig. 2 (b). In Figs. 2 (a) and (b), we can see that the HSII networks learn the corrupted or crooked characters by improving weight vectors. In table 1, we show the number of hidden units required to represent each categories. In the HSII, fewer hidden units are required in every category than in the RCE.

Table 1. The number of hidden units which were required to represent each categories.

Category	HSII	RCE
0	3	7
1	2	3
2	1	3
3	3	3
4	1	2
5	3	4
6	4	6
7	2	3
8	3	6
9	1	4
M	1	2
T	1	2
S	6	10
Total	31	55

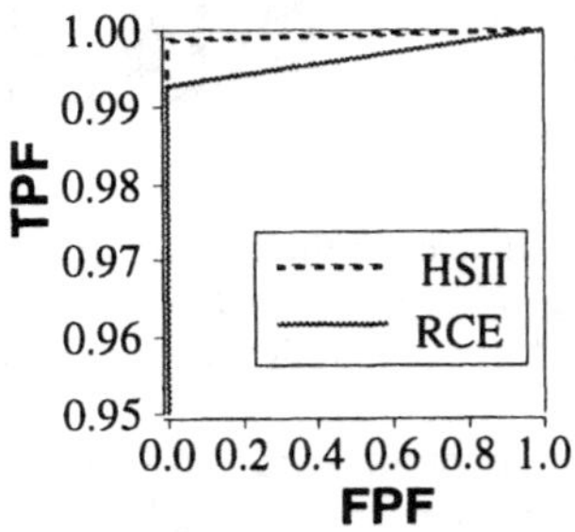

Fig. 3 ROC curves which evaluate the performances of networks. Dotted and gray lines denote ROC curves in the case of the HSII and RCE, respectively.

To compare the performances of networks, the receiver operating characteristic (ROC) analysis [5] is available. The ROC is an analytical method generally utilized to evaluate the performance of a system. ROC analyzes overlapped distributions of positive and negative outputs of a system. A positive output means an output to be fired and a negative output means an output not to be fired. When the distributions of both positive and negative outputs overlap, a threshold can be made at an arbitrary point in the overlapped distribution area for judgment. Thresholding outputs makes true positive and false positive outputs. The plot of the corresponding true-positive fraction (TPF) versus false-positive fraction (FPF) for each threshold constructs a ROC curve. The area under the curve called "Az" can be read as a performance index of the system using ROC analysis. In general, the higher the Az the better the performance.

To evaluate the performances, we tested networks trained by the HSII and the RCE using a 1,200-element testing set extracted from 46 sheets of unknown X-ray films. In Fig. 3, dotted and gray lines denote ROC curves of networks trained by the HSII and the RCE, respectively. In this figure, the axis of TPF is expanded to clearly show the difference of curves. The Az of the HSII is larger than the RCE. From this result, we can show that the performance of the network trained by the HSII is superior to by the RCE.

We can determine an optimal threshold from ROC curves shown in Fig. 3. In this application, it is not expected to mis-recognize any characters. That is to say, FPF must be zero. The point in which FPF is zero and TPF is maximum is obtained when the threshold is 0.95 in the HSII. In the case of the RCE, the point is obtained when the threshold is 0.67. These values were used as optimal threshold in our experiments.

When the threshold is made at a certain point, the network has the following three states; (i) having only one fired output unit and it's to be fired, (ii) having only one fired output unit but it's not to be fired, and (iii) having no, two, or more fired output units. The first and second cases can be regarded as a correct recognition and a mis-recognition, respectively. The last one is defined as a rejection. With regard to the network trained by the HSII when the threshold was made at 0.95, a recognition rate, a mis-recognition rate, and a rejection rate are 99.4%, 0.0%, and 0.6%, respectively. On the other hand, in the case of the RCE when the threshold was at 0.67, a recognition rate, a mis-recognition rate, and a rejection rate are 97.8%, 0.0%, and 2.2%, respectively. From this results, we can also show the network trained by the HSII has a higher generalization ability in spite of fewer hidden units than by the RCE.

4 Conclusions

We have proposed a learning algorithm, called HSII method. In the procedure of the HSII, the number of hidden units are automatically determined, and the interconnection weights between the input and hidden units are optimized by back propagation of errors calculated in the output layer. We can summarized the advantages of the HSII are; (1) optimization of network size, and (2) network parameters, (3) reduction of learning time, and (4) generalization against untrained inputs.

We demonstrated a successful application of the HSII to character recognition on X-ray films. The network was trained on 2,250 training examples consisting of the numerals 0-9 and three alphabets, T, M, and S. It achieved a recognition rate of 99.4% and a mis-recognition rate of 0.0% on a 1,200-element testing set. We also compared the HSII to the RCE by the ROC analysis. By the comparison, we showed the network trained by the HSII has a higher generalization ability in spite of fewer hidden units than by the RCE.

Acknowledgments

We gratefully acknowledge the constructive suggestions of Dr. Shih-Chung B. Lo and Mr. Jyh-Shyan Lin.

References

[1] D. E. Rumelhart, G. E. Hinton, and R. J. Williams, "Learning internal representations by error propagation," in *Parallel Distributed Processing*, ed. D. E. Rumelhart and J. L. McClelland, pp. 318-362 (Cambridge, MIT Press, 1986).

[2] A. Hasegawa, W. Zhang, K. Itoh, and Y. Ichioka, "Neural Network-based Image Processing on Human Corneal Endothelial Micrograms," in *Proc. Soc. Photo-Opt. Instrum. Eng.*, Wave Propagation and Scattering in Varied Media II, 1558, pp. 414-421 (San Diego, Carifornia, 1991).

[3] Y. LeCun, B. Boser, J. S. Denker, D. Henderson, R. E. Howard, W. Hubbard, and L. D. Jackel, "Backpropagation Applied to Handwritten Zip Code Recognition," *Neural Comp.*, 1, pp. 541-551 (1989).

[4] D. L. Reilly, L. N. Cooper, and C. Elbaum, "A Neural Model for Category Learning," *Biol. Cybern.*, 45, pp. 35-41 (1982).

[5] R. C. Eberhart and R. W. Dobbins, "Receiver Operating Charactericstic Curves" in *Neural Network PC Tools*, pp. 169-174 (New York, Academic Press).

Diagnosis of Technical Equipment using a Bayesian Neural Network

Anders Holst and Anders Lansner

SANS—Studies of Artificial Neural Systems
Department of Numerical Analysis and Computing Science
Royal Institute of Technology, S-100 44 Stockholm, Sweden
aho@sans.kth.se, ala@sans.kth.se

Abstract

The Bayesian neural network implements an optimal Bayesian classifier in the case of independent evidence. If the evidence is not independent, a hidden layer with an independent representation can be used in the network, to overcome this limitation on the evidence. This Bayesian neural network has been tested on a realistic task of classifying faults in a telephone exchange computer, with very good results. The results are compared to those of a backpropagation neural network on the same data. Also presented is an explanatory mechanism for finding the primary causes for a classification.

1 Introduction

Classification or diagnosis is a very general and important area, with many useful applications. Various methods for automatic diagnosis have been tried over the years. One approach is to use neural networks for such applications.

Here we will consider a Bayesian neural network. It will be used for classification of faults in a technical system, a telephone exchange computer, consisting of of number of circuit cards. When a fault is detected in the system, a binary error vector is generated. The network is used find out which card is most likely to be malfunctioning, given the error vector.

2 The Bayesian Neural Network

The neural network model used here is a one-layer fully-connected Bayesian neural network [6]. In this network the activities of units are related to probabilities of stochastic events. The model is based on Bayes rule for conditioned probabilities. According to this rule the probability of an event q, *e. g.* a class, given a set of independent events $A = \{x_i, x_j, x_k \ldots\}$, *e. g.* features, can be calculated as:

$$P(q \mid x_i, x_j, x_k \ldots) = P(q)\frac{P(x_i, x_j, x_k \ldots \mid q)}{P(x_i, x_j, x_k \ldots)} = P(q) \prod_{x_i \in A} \frac{P(x_i \mid q)}{P(x_i)}. \quad (1)$$

By taking the logarithm of (1), it may be rewritten as a sum [7]:

$$\log(P(q \mid x_i, x_j, x_k \ldots)) = \log(P(q)) + \sum_i \log\left(\frac{P(q,i)}{P(q)P(i)}\right) o_i, \quad (2)$$

where $o_i = 1$ if $x_i \in A$, and 0 otherwise.

Now we can identify events (features x_i as well as classes q) with units in a neural network. Thus o_i is the output from unit i, and the biases β_q and weights w_{qi} are chosen as:

$$\beta_q = \log(P(q)), \quad (3)$$

$$w_{qi} = \log\left(\frac{P(q, x_i)}{P(q)P(x_i)}\right). \quad (4)$$

The left hand side of (2) is called the *support* of unit q. By using an exponential transfer function of the support, we get the posterior probability of the event.

To be able to explicitly state that an event has not occurred, we use complementary coding, *i. e.* each event is represented with one positive and one negative unit in the network. To use the network, the units corresponding to observed features are stimulated, the activity spreads through the weights to the class units, and the resulting activity is interpreted as the probability of the different classes given the observed features. If the class with the highest probability is chosen, we have an optimal Bayesian classifier [1].

Training of the network is done in one pass over the training set, during which the probabilities $P(x_i)$ and $P(q, x_j)$ are estimated from the training examples. These estimations are then used to set biases and weights according to (3) and (4). Correlated units thus get positive connections between them, anti-correlated units get negative connections, and uncorrelated units get zero connections.

The details of the model and its mathematical background are more thoroughly examined in [6, 2]. Similar neural network models, based on probabilities, are treated in [4, 8].

3 A Multi-Layer Bayesian Neural Network

The Bayesian model assumes independence between different features. If this assumption is not fulfilled, the estimation of probabilities in the network will degrade. One way to solve this is to create complex units, *i. e.* units that combine the information from a set of primary units [5]. Chosen in the right way they can be made to constitute a hidden layer with independent features, upon which the ordinary Bayesian model can work.

In the network used here, these complex units are grouped into complex columns, *i. e.* groups with one unit for each possible combination of outcomes of some primary features. For example, if the two primary features $A = \{a, \bar{a}\}$ and $B = \{b, \bar{b}\}$ are not independent, they are merged into a complex column $AB = \{ab, a\bar{b}, \bar{a}b, \bar{a}\bar{b}\}$. If we manage to partition the primary features into independent groups, we can create a complex column from each such group. This will then solve the problem with dependencies in the Bayesian model.

In general it is a hard problem to find the independent partition of lowest order. However, a heuristic method that has shown itself very useful, is to merge two features into a complex column if some measure of the correlation between them exceeds some threshold. The merging of columns can continue for a few passes, *i. e.* if two columns created in the first pass are too much correlated, they are merged into a still higher order column. The measure of correlation that is used here, is the covariance between two units i and j, $P(x_i, x_j) - P(x_i)P(x_j)$. This method has in tests efficiently reduced the amount of correlation in the network, often sufficiently after only one or two passes (*i. e.* at most fourth order columns).

4 The Diagnosis Application

This Bayesian model has been tested on a realistic task: diagnosis of telephone exchange computers. In this case, one malfunctioning circuit card out of 36 is to be identified from an error vector of 122 bits. The training set, consisting of 442 real examples, is not completely unambiguous (malfunctioning of different cards can give rise to the same symptoms), and in general there are only small differences between the vectors from different fault classes. In addition to the training set, a test set of 112 real examples is used.

Classification of these data is tried with both the one-layer and the multi-layer versions of the Bayesian neural network. The one-layer network is as described above, with complementary coding of the error bits, *i. e.* 244 input units. The multi-layer Bayesian network is constructed by merging all pairs of primary features (error bits) with a covariance larger than 0.01 into complex columns. This corresponds to one pass of the above algorithm, which produces at most second order columns. We also use another multi-layer Bayesian network where the threshold for the covariance is 0.001, and columns are created in three passes (denoted "Trimmed Bayes" in the figures and the table). This produces columns of higher order, which models the distribution of the training data more accurately.

For comparison, a backpropagation neural network is also used. It has one hidden layer of 30 units, and is trained with 170 epochs. Further, there is a conventional diagnosis program for the telephone exchange computer system. This program manages to classify about half of the examples (in both training and test set) correctly.

Table 1 summarizes the results with the neural networks. The fraction of correct answers among the first, second, and third alternatives from the networks are shown. The same data is presented in figure 1. All the neural networks achieved better results than the conventional method. The

Table 1: Classification Results. The networks and the conventional method are tested with both the training and the test set. The percent of correct answers among the first, second and third alternatives from the networks are shown.

Tested on training set (results in %)

	Conv.	Backprop	One-layer	Multi-layer	Trimmed
First	47.7	91.2	75.1	80.8	89.1
Second	7.2	3.4	15.4	11.5	7.0
Third	0.7	0.5	4.3	3.8	1.1
Rest	44.3	5.0	5.2	3.8	2.7

Tested on test set (results in %)

	Conv.	Backprop	One-layer	Multi-layer	Trimmed
First	50.0	67.0	66.1	72.3	68.8
Second	5.3	8.9	12.5	9.8	8.9
Third	0.0	0.9	8.0	5.3	4.5
Rest	44.6	23.2	13.4	12.5	17.9

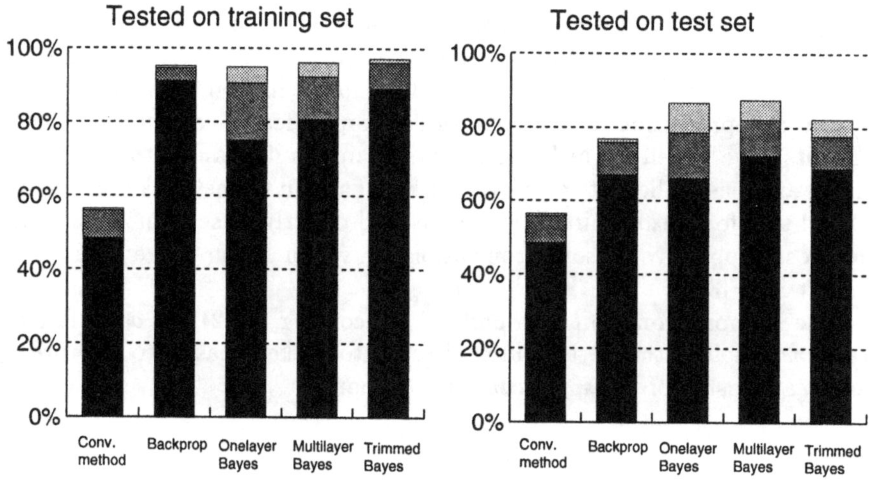

Figure 1: Classification Results. Black represents the fraction with correct first alternative, dark grey correct second alternative and light grey correct third alternative.

results of the Bayesian networks are comparable to that of backpropagation. Compared to the one-layer and first multi-layer Bayesian network, backpropagation seems a little better when tested on the training data, if only the fraction of correct first alternatives is considered. However, for generalization (*i. e.* results with the test set) the different networks give approximately the same fraction of correct first alternative. If the second and third alternatives are counted as well, the Bayesian network generally leaves a smaller fraction in the "Rest", specifically for generalization but also on the training data. In the second multi-layer Bayesian network, for which the threshold of the correlation for creating a complex column is decreased, and merging of columns is done in a few passes, we get the results in the column "Trimmed Bayes". In this column the Bayesian network is seen to have nearly the same results as the backpropagation network, but the generalization capability has decreased again.

5 An Explanatory Mechanism

An important feature of a classification system, is the capability to give some explanation of the systems result. If the system is not able to account for its conclusions, it will be harder for people to trust it. In rule based systems it is possible to achieve simple explanations by showing the user which rules were used for the deduction. In contrast, this is often considered as a problem in neural networks, which do not deal with explicit rules.

The units in the Bayesian neural network has an interpretation as stochastic events. This is true also for the units in the hidden layer, which can be seen as representing joint events (or as complex feature detectors). That all units have a well defined interpretation, makes it possible to achieve a simple analysis of the state of the network. Signals in the network are closely related to information gains, and can be used directly to see which features are most supportive of some conclusion, or which features are against a certain class [3].

The support from unit i to unit q is, according to (2), $w_{qi}o_i$. After a transformation of this to the interval $[0,1]$, to make it easier to interpret, we get a measure of the ***significance*** for q from i:

$$g_{qi} = \frac{w_{qi}o_i}{-\beta_q} \tag{5}$$

A significance of zero means that the unit i does not tell anything about q, whereas a significance of one means that the event q is made completely certain by i. If i contributes evidence against q, the significance of i for the negation of q is used, rather than treating negative values of the significance.

In the case of faults above, this means that it is possible to stimulate the network with an error bit vector, and get a list of the most probable erroneous circuit cards. Then the explanatory mechanism is used to find what bits in the error vector are the strongest reasons for (or against) believing that a card is the failing one.

6 Conclusions

Diagnosis of complicated technical systems that are hard to analyze, is a good example of an area in which neural networks can be useful. This study has shown good results when applying neural networks to diagnosis of a telephone exchange computer.

The performance in this application of a Bayesian neural network and of a backpropagation neural network are comparable, but the Bayesian network tends to be slightly better at generalization. This might be due to the fact that the Bayesian model is based on probabilities, instead of just trying to decrease the error in the training set.

There are further advantages of the Bayesian model over backpropagation. The training time is much shorter for the Bayesian network, since training is done in one pass, or a few passes for the multi-layer version. For example on a DEC-Station 3100 it takes two to three minutes for the Bayesian network, compared to about two hours for backpropagation. Also, with the interpretation of units as stochastic events, it is possible to have a kind of explanatory mechanism, which motivates a hypothesis in terms of which features support it most. This is usually not possible in a backpropagation network, where the hidden units and the weights have no simple interpretations but are merely selected to solve the problem.

The Bayesian model has a large potential for further development. There are plenty of ways to generalize the concept of complex columns, and also other ways to select which primary events to merge into complex columns. But already this network model has shown itself a powerful and general tool for classification and diagnosis.

Acknowledgments

We are grateful to P. Hammarlund for valuable comments on the manuscript, and to M. Boda and H. Brandt for useful comments and discussion. This work was supported by Ellemtel Telecommunication Systems Laboratories (Ellemtel Utvecklings AB).

References

[1] R. O. Duda and P. E. Hart, *Pattern Classification and Scene Analysis.* Wiley, New York, 1973.

[2] A. Holst and A. Lansner, "A Bayesian neural network with extensions," Tech. Rep. TRITA-NA-P9325, Dept. of Numerical Analysis and Computing Science, Royal Institute of Technology, Stockholm, Sweden, 1993.

[3] I. Kononenko, "Bayesian neural network-based expert system shell," *Int. J. Neural Networks*, vol. 2, no. 1, pp. 43–47, 1991.

[4] I. Kononenko, "Bayesian neural networks," *Biol. Cybernetics*, vol. 61, pp. 361–370, 1989.

[5] A. Lansner and Ö. Ekeberg, "An associative network solving the "4-Bit ADDER problem"," in *Proceedings of the IEEE First Annual International Conference on Neural Networks*, vol. 2, (San Diego, USA), pp. 549–556, June 1987.

[6] A. Lansner and Ö. Ekeberg, "A one-layer feedback, artificial neural network with a Bayesian learning rule," *Int. J. Neural Systems*, vol. 1, no. 1, pp. 77–87, 1989.

[7] M. L. Minsky and S. A. Papert, *Perceptrons.* MIT Press, 1988.

[8] P. Orponen, P. Floréen, P. Myllymäki, and H. Tirri, "A neural implementation of conceptual hierarchies with Bayesian reasoning," tech. rep., Department of Computer Science, University of Helsinki, Finland, 1990.

A Connectionist Approach for Incorporating Continuous Code Churn into Software Reliability Growth Models

N. Karunanithi

Room 2E-378, Bellcore
445, South Street, Morristown, NJ 07960
karun@faline.bellcore.com

Abstract

A large number of execution-time based reliability growth models have been proposed for estimating reliability of software systems. One of the key assumptions made in almost all of the models is that the complete code for the system is available before testing starts and that the code remains frozen during testing. However, this assumption is often violated in large software projects because usually the code is developed in parts. This paper demonstrates the applicability of the neural network approach to the problem of developing an extended software reliability growth model in the face of continuous code churn. In this preliminary study, neural network reliability models with and without the code churn information are compared using a data set from a large telecommunication system. The results suggest that the neural network model with the code churn information is capable of providing a more accurate prediction of future faults than the model without the code churn information.

1 Introduction

A large number of dynamic reliability models for have been proposed for estimating reliability growth of software systems. These analytic models describe the failure process as a function of execution time (or calendar time) and are based on many simplifying assumptions [8]. One of the key assumptions made by many of the existing analytic models is that the code size remains unchanged during testing. This assumption may not be valid for most large software systems because the program undergoes change during development.

Evolving programs can introduce a variety of complications for both software engineers and the software reliability estimate. For software engineers, an evolving program becomes a moving target during testing. Furthermore, the testing team cannot have uniform confidance on all parts of an evolving program because the code which is included early on would be exercised more often than the parts that are included later. From software reliability estimation point of view, the program evolution introduces complications in efforts such as data collection, modeling, analysis and interpretation. Thus, in the face of program evolution the existing execution time based software reliability growth models may not be able to provide a trustworthy estimate without proper extensions.

In order to deal with a continuously evolving code (or "code churn") Dalal and McIntosh [2] proposed a simple extension to a Poisson process model and empirically demonstrated that adding the code churn information can improve the fit of the model. This paper demonstrates the applicability of the neural network approach to the problem of modeling software reliability growth in the face of continuous code churn. In this preliminary empirical study, neural network reliability models with and without the code churn information are compared using a data set from a large telecommunication system. The results suggest that the neural network model with the code churn information is capable of providing a more accurate prediction of future faults than the model without the code churn information.

2 The Data Set

The data set used for this study is from a large telecommunication system consisting of approximately 7 million non-commentary source lines [2]. The particular release for which the reliability model is applied had around 400,000 new or changed non-commentary source lines (NCNCSL). Since the testing was performed in a highly distributed environment the time was measured in "staff days". The "staff day" metric represents the amount of time a tester actually spends every day on testing the current release. Figure 1 illustrates the continuous code churn as a function of staff days. The vertical (dashed) line indicates the date on which the system was frozen. After the "soft freeze", no functionality was added to the system and subsequent changes reflect only fault fixes. Though several subsystems were delivered at the start of the testing, only a subset of them were actually tested. This

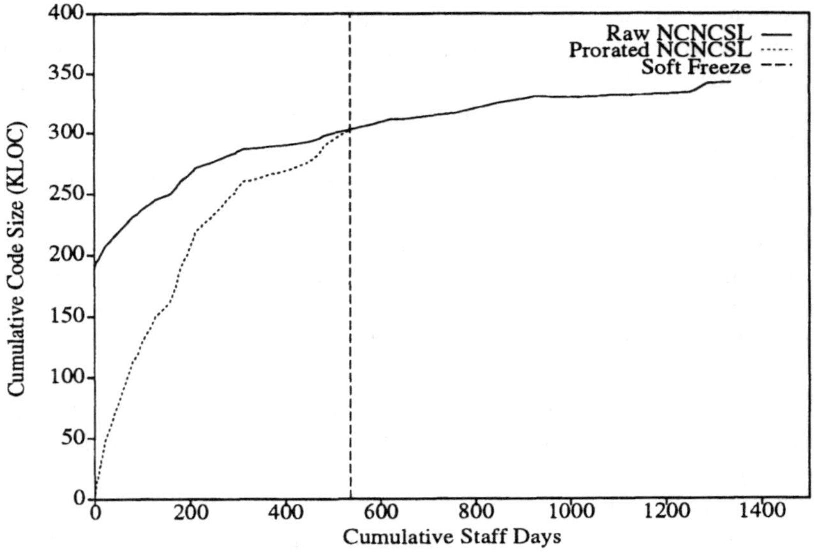

Figure 1: Code Churn Vs. Cumulative Staff Days.

necessitated a proper adjustment to the initial NCNCSL. The dotted line in Figure 1 shows the adjusted code size.

3 Model Development

3.1 Dalal and McIntosh Model

The basic model of Dalal and Mallows [1] on which the extended model is built is given by

$$\mu(t_i) = \alpha_i(1 - e^{-\beta(t_i - t_{i-1})})$$

where $\mu(t_i)$ is the mean number of faults found at time t_i, α_i is the number of faults at the beginning of the i-th interval and β is the rate parameter. The number of remaining faults at time t_i is equal to $\alpha_i e^{-\beta(t_i - t_{i-1})}$. The extended model which incorporates code churn is given by,

$$\alpha_{i+1} = g(\alpha_i e^{-\beta(t_i - t_{i-1})}, c_i, \theta) \qquad i = 1, \ldots, n$$

where $g(x, c, \theta)$ is a general (yet unspecified) function and c_i is the size of the code added at the end of i-th interval. If there was no code added in an interval then the corresponding $c_i = 0$. Typically, the extended model can also be expressed as

$$g(x, c, \theta) = x + g_1(c, \theta) + g_2(x, c, \theta)$$

with $g_1 = g_2 = 0$ whenever $c = 0$. This formulation can be interpreted as follows.

The number of faults in the code after the code churn =
{ the number of faults immediately before the code churn } +
{ the number of faults in newly delivered code } +
{ the number of faults in the code because of the interactions between the two sets of faults}.

As a first step approximation, it is assumed that the function $g_2(x, c, \theta) = 0$. This is equivalent to saying that there is no additional faults due to interaction of the existing code and the newly added code. Next, they assume that the structure of g_1 is an identity function and that c is equal to NCNCSL at i-th delivery (i.e., $g_1(c, \theta) = \theta g_1(c)$). Finally, they simplify the model by assuming that the additional faults in the newly added code is proportional the size of the code. Thus, the resulting model is given by,

$$g(\alpha_i e^{-\beta(t_i - t_{i-1})}, c_i, \theta) = \alpha_{i+1} = \alpha_i e^{-\beta(t_i - t_{i-1})} + \theta c_i$$

where α_{i+1} is the number of faults in the system at the beginning of $(i+1)$st interval. The corresponding expression for the cumulative fault, $M(t_i)$, is given by,

$$M(t_i) = M(t_{i-1}) + \alpha_i(1 - e^{-\beta(t_i - t_{i-1})})$$

where $M(t_{i-1})$ is the cumulative faults found at the end of the $(i-1)$st interval. Note that the above expression has the structure of an autoregressive process. One of the important claims made by Dalal et. al., [1] is that the fit of the extended model is considerably superior than that of the basic model.

3.2 The Neural Network Approach

Applicability of neural network models to software reliability growth prediction has been demonstrated by Karunanithi et al. [5, 6, 7]. Two important conclusions of these studies are: i) the neural network approach is a "black box" approach (i.e., the neural networks are capable of developing an appropriate model for the failure process from the training data) and ii) the modified Jordan style [4] networks with "Teacher Forced" training are capable of providing a more accurate predictions than the feed-forward networks.

The network models used in the previous research[5, 6, 7] had the accumulated time (t_i) as the free variable and the cumulative faults ($M(t_i)$) as the dependent variable. From the modeling point, they are analogous to the traditional execution time based software reliability growth models. However, the networks used in the present study have an additional input

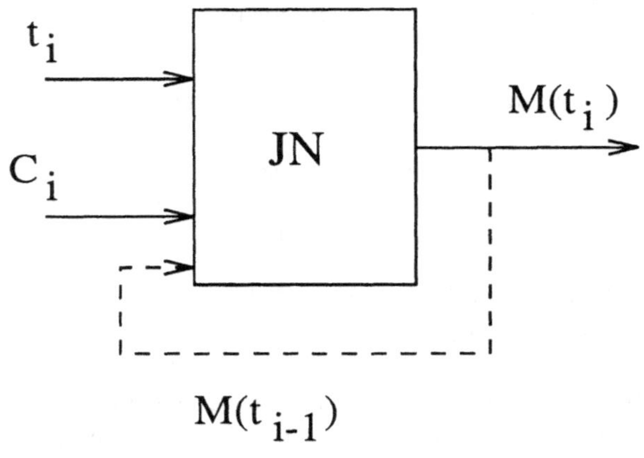

Figure 2: A Modified Jordan Style Neural Network.

(C_i) for the cumulative code churn. Figure 2 represents an abstract model of the modified Jordan style network used in this study. The dotted line in Figure 2 represents the feed-back from the output $(M(t_{i-1}))$ required for realizing the Jordan style network. For simplicity, it is assumed that the network operates only in discrete time-steps. The "box" (JN) represents the neural network.

4 Prediction Results

Even if the fit of a model is good, that does not guarantee that its predictions will always be accurate. A good model also should provide accurate predictions of future faults. In order to assess the predictive capability of the model we used two extreme prediction horizons: the *next-step prediction* (NSP), for the cumulative faults at the end of the next time step; and the *end-point prediction* (EPP), for the cumulative faults at the end of the test phase. When an extended model like the one developed in this paper is used for predicting future events, one has to know not only the execution time corresponding to a future day but also a precise information about the code churn. We solved this issue by using the size of the code corresponding to the final value of the code churn in the training data to all future predictions (i.e., the value of the code size of the last point in the training data was considered as the size of the final code). Note that this issue does not arise if we do not use the code churn information.

The neural networks cannot predict well without sufficient training data. This is analogous to using insufficient data to estimate the parameters of a stochastic model. In our prediction experiment, the size of the training set was gradually increased from a minimal set consisting of all data points before the "soft freeze" up to a set with all but the last point in the failure history. In order to gauge the predictive quality of the neural network model we used the average relative error (ARE) used by Malaiya et al. [9]. The average relative error measure is defined as

$$ARE = \frac{1}{n-k-1} \sum_{i=k+1}^{n-1} \left| \frac{M(t_i) - \hat{M}(t_i)}{M(t_i)} \right|$$

where n is the number of points in data set, k is the number of points in the training set corresponding to the soft freeze and $n-k-1$ is the prediction window. ARE provides a summary of how well the model predicts across a window of the future failure history. The predictive performance of the neural network models for the two extreme prediction horizons are summarized in Table 1. The "Mean" and "SD" represent the mean and the standard deviation of the ARE measure over 50 trials. These results suggest that the network with the code churn information is capable of providing a more accurate prediction than the network without the code churn information.

Neural Network	NSP		EPP	
Model	Mean	SD	Mean	SD
Without Code Churn	3.834	1.883	26.342	8.417
With Code Churn	2.955	1.803	17.27	5.837

Table 1: A Summary of Prediction Results in Terms of ARE.

5 Conclusion

We demonstrated that one can easily extend the execution time based neural network framework to incorporate the code churn information. Our preliminary results suggest that incorporating the code churn information can help both the fit as well as the predictive accuracy of the neural network models. However, we do note that the results presented here are preliminary and they have yet to be compared with the results of the existing models. This will be explored in future.

Often, one is interested in not only predicting the cumulative faults but also other quantities such as the rate of occurrence of failure, mean time to failure etc. The extended neural network model developed in this study represent only the cumulative faults. However, if one is interested in the failure rate expression, it is straightforward to derive it from the expression for the cumulative faults using the method outlined in [7].

References

[1] Dalal, S. R. and Mallows, C. L., "Some Graphical Aids for Deciding When to Stop Testing Software", *IEEE J. Selected Areas in Communications.*, Vol. 8, No. 2, pp. 169-175, 1990.

[2] Dalal, S. R. and McIntosh, A. A., "Reliability Modeling and When to Stop Testing for Large Software Systems in the Presence of Code Churn: Analysis and Results for TIRKS Release 16.0", *Bellcore, TM-ARH-021705*, Aug. 1992.

[3] Fahlman, S. E. and Lebiere, C., "The Cascaded-Correlation Learning Architecture", School of Computer Science, Carnegie Mellone University, Tech. Rep. CMU-CS-90-100, Feb. 1990.

[4] Jordan, M. I., "Attractor Dynamics and Parallelism in a Connectionist Sequential Machine", *Proc. of the 8th Annual Conf. of the Cog. Science*, 1986, pp. 531-546.

[5] Karunanithi, N., Whitley, D. and Malaiya, Y. K., "Prediction of Software Reliability Using Connectionist Models", *IEEE Trans. on Software Eng.*, Vol. 18, No. 7, pp. 563-574, July 1992.

[6] Karunanithi, N., Whitley, D. and Malaiya, Y. K., "Using Neural Networks in Reliability Prediction", *IEEE Software*, Vol. 9, No. 4, pp. 53-59, July 1992.

[7] Karunanithi, N. and Malaiya, Y. K., "The Scaling Problem in Neural Networks for Software Reliability Prediction", *Proc. 1992 Int. Symp. on Soft. Rel. Eng.*, pp. 76-82, Oct. 1992.

[8] Littlewood, B., "Theories of Software Reliability: How Good Are They and How Can They Be Improved?", *IEEE Trans. on Software Eng.*, Vol. SE-6, No. 5, pp. 489-500, Sep. 1980.

[9] Malaiya, Y. K., Karunanithi, N. and Verma, P., "Predictability of Software Reliability Models", *IEEE Trans. Reliability*, Vol. 41, No. 4, pp. 539-546, Dec. 1992.

Real-Time Identification of Language from Raw Speech Waveforms

Stan C. Kwasny, Barry L. Kalman, A. Maynard Engebretson, and Weilan Wu

Department of Computer Science
Washington University
St. Louis, Missouri 63130 USA
sck@cs.wustl.edu
barry@cs.wustl.edu
ame@cs.wustl.edu
wwl@cs.wustl.edu

Abstract

Real-time language identification is a problem well-suited to neural network solution. By limiting the amount of speech processing at the frontend this can be achieved to a surprising degree. In our experiments we have developed a recurrent neural network which has been trained and tested using examples from English and French. The recurrent network produces a series of activation patterns (one outcome for each cycle) which can be seen as votes for or against each language. These votes are summed over an interval of speech and the segment is classified.

1 Introduction

Unquestionably speech will play an increasingly important role in telecommunications. Technology will continue to bridge vast distances bringing people from diverse cultures and languages together electronically. At the front end of any multi-language speech component will be the ability to identify which language the speaker is speaking in order to translate among languages, inform or instruct speakers, and otherwise communicate with them.

Humans perceive spoken language on many levels. A variety of judgements about features of speech are constantly being made by a listener. Listeners unconsciously notice many things about speech tone of voice, style, pace, gender of the speaker, accent, degree of excitement, the identity of the speaker, etc. These high-level features are not easily computed from speech signals nor are they consciously contemplated under ordinary circumstances and yet may dramatically influence our interaction with the speaker. Further observe that spoken communication among humans utilizes a myriad of devices to convey messages including frequency, pitch, and sequencing, as well as other prosodic and durational properties measurable in the signal. If a muffled conversation is overheard through a thin wall, it may possible to identify the language even when not understanding exactly what was spoken. The complexity of interactions among these properties in the speech signal are impossible to capture in any simplistic model necessitating the use of models such as Hidden Markov Models and neural networks.

1.1 Related Work

Interest in language identification has increased dramatically in recent years. Leading much of this research has been the Oregon Graduate Institute (OGI) who have collected a database of phone speech in ten languages described in [10] and have built a system that distinguishes among those languages [9].

In the past, there have been studies that demonstrate the existence of statistically significant differences among spoken languages at the acoustic level [5] [2] and also at the level of phonetic features [3] [8]. Abe et al [1] have considered some of the differences in automatically converting a speaker's voice from one language into another. Since these differences are measurable at the low end of the speech chain, then surely it must be possible to exploit those differences to build a model that emulates the human ability to correctly discriminate among languages.

1.2 The Problem

While speech understanding research has focused primarily on extracting "meaning" from speech, it is clear that there are many other ways humans process speech. Most of the high-level features mentioned above cannot be tied to any particular, conventional set of phonetic or acoustic features of the speech. Instead, they appear to be related to distributional patterns or statistical aggregates of the speech waveform.

Real-time language identification is a problem well-suited to neural network solution. By limiting the amount of speech processing at the frontend, this can be achieved to a surprising degree. In our preliminary studies, reported earlier, using a feedforward network [7] and using a recurrent neural network [11], we have shown that classification can be conducted in real time for two speakers, male and female. We have assumed that it is only necessary for the model to see relatively raw speech waveforms, represented as sampled frequency bands over time. This contrasts with attempts based on explicit phonetic identification as well as a variety of other intermediate-level structuring typically found in speech understanding and recognition systems.

In this paper we expand our investigation by examining performance on telephone speech. We report on preliminary results using the OGI multi-language database of telephone speech.

2 Data Collection

The OGI database contains recordings of telephone quality speech in ten different languages recorded at 8kHz. We randomly selected samples from two French speakers and two English speakers in which the speaker is conveying a story. Each story takes approximately 50 to 60 seconds and these were cut into multiple 10-second segments to yield training and test data. A total of 16 segments were randomly selected. One segment was randomly selected from each of the four speakers for test purposes and 12 segments were selected for training with five French and seven English segments being selected in this way.

Within each 10 second sample we passed a fixed-duration (1.2 sec) window over the signal moving it forward in 95 ms. increments to form 95 overlapping segments of speech. These segments were processed into patterns for the neural network training and testing by applying five band-pass filters which divided the sampled speech into the bands 67Hz – 133Hz, 133Hz – 266Hz, 266Hz – 533Hz, 533Hz – 1066Hz, and 1066Hz – 2133Hz. The output from each was low-pass filtered and decimated by a factor of 200 to provide real-valued activation patterns for the network. These were presented to the recurrent network in the proper time sequence.

3 Network Design and Training

Our network architecture is based on a simple recurrent network [4], in which the activation pattern of the hidden units is copied back as additional input unit activations. In our version, the input units and copyback units are also connected directly to the output allowing those connections to learn the linear portion of the required mapping. The input patterns require 240 units for each 1.2 second segment, while only four hidden units (and consequently four copyback units) were required to give optimal generalization. There were two output units, one for English and one for French.

Each 10 seconds produces 95 outcomes since the overlapping window fits into the 10-second interval 95 times. We treat these as votes with a majority required to force the classification of the segment. However, some initial number of votes (in these experiments, 11) are not counted since they depend too closely on the initial pattern chosen for the copyback units. This results in 84 counted votes from each segment.

For training to succeed, we needed to apply singular value decomposition (svd) to the input patterns essentially re-orienting the data to maximize orthogonality among the input unit activations [6]. We have found this particular technique valuable in a variety of problems.

With this, training succeeded. As each window is moved across the signal, the network decides if it contains an English or a French segment based partially on the new data entering the window and partially on the copyback pattern from the previous vote. The voting strategy permits the network to classify incorrectly for short intervals within the segment as long as the overall classification is correct.

4 Results

The results of these experiments are shown in Table 1.

Table 1: Voting on Test Segments

Language	**Sample**	**Votes (out of 84)**
English	1	81
English	2	65
French	1	59
French	2	84

All training segments trained with a perfect score of 84 votes. All test segments, as shown in the table, were classified correctly with the lowest total earned for the first French sample which still won by a convincing 59 to 25 majority. In examining the hidden unit patterns, they cluster readily into two very disjoint groups with a small number of minor subgroupings within each. It is clear from examining these cluster patterns that excellent separation is occurring.

5 Conclusions

Although our results are preliminary, we have reason to be optimistic about the techniques employed. Analysis of our voting scheme as a discrete, binomial experiment indicates that we can trade off network performance vs. number of votes. This implies that a more difficult discrimination is possible by allowing the network to generate more votes.

5.1 Plurality Voting

Our technique is expected to scale up as more categories are added. This can be seen by analyzing the length of a voting sequence for a multiple outcome classification as a function of estimated probability of each outcome. The estimation algorithm uses the binomial distribution to compare the probability that the target outcome receives a majority of the votes and uses the multinomial distribution to compute the probability that the target outcome receives a plurality but not a majority of the votes.

¿From the binomial distribution we get

$$P_M = \sum_{k=\lceil V/2 \rceil+1}^{V} p_1^k q^{V-k} \binom{V}{k} \tag{1}$$

where we assume the following inputs:

$$\begin{aligned} M &= \text{number of categories} \\ V &= \text{number of votes} \\ p_i &= \text{probability of correctly predicting category } i, 1 \le i \le M. \end{aligned}$$

and, without loss of generality, we assume that outcome category 1 is the target. For the majority portion of the probability estimate, define $q = (1 - p_1)$ and P_M as the probability that category 1 gets a majority of the votes.

For the plurality probability P_L we use the multinomial distribution. The plurality result is

$$P_L = \left(\sum_{k_1=n_1}^{N_1} \sum_{k_2=n_2}^{N_2} \cdots \sum_{k_{M-1}=n_{M-1}}^{N_{M-1}} \left(\prod_{I=1}^{M} \frac{p_i^{k_i}}{k_i!} \right) V! \right) \tag{2}$$

For this result to hold, we must determine the correct bounds for each sum of the multiple summation. The sum over the first category must go from $k_1 = \lceil V/M \rceil + 1$ to $\lceil V/2 \rceil$. None of the rest of the summation indices can exceed $k_1 - 1$ or the number of votes remaining. None of the rest of the summation indices can fall below zero and $\sum_{i=1}^{M} k_i = V$. Let n_i be the minimum of index k_i and N_i be its maximum.

Clearly $n_1 = \lceil V/M \rceil + 1$ and $N_1 = \lceil V/2 \rceil$. For $2 <= i <= M - 1$ we get

$$n_i = MAX(0, V - (\sum_{j=1}^{i-1} k_j + (M - i)(k_1 - 1))) \tag{3}$$

and

$$N_i = MIN((k_i - 1), V - (\sum_{j=1}^{i-1} k_j)) \tag{4}$$

with

$$k_M = V - (\sum_{j=1}^{M-1} k_j) \tag{5}$$

6 Future Work

¿From these preliminary results and our extrapolation showing how the number of votes relates to the performance, we believe these results will generalize well. Others have studied how to combine different estimators while we view our voting scheme as a way to leverage a single estimator on a problem that permits multiple votes but requires a single classification. Obviously more work is required to strengthen that case. Such experiments are continuing.

Acknowledgements

We are grateful to OGI for providing the data on which these results are based.

References

[1] M. Abe and K. Shikano "Statistical analysis of bilingual speaker's speech for cross-language voice conversion," *Journal of the Acoustic Society of America,* vol. 90, pp. 76–82, 1991.

[2] K. Atkinson "Language identification from nonsegmental cues," *Journal of the Acoustic Society of America,* vol. 44, p. 378(A), 1968.

[3] P.B. Denes "On the statistics of spoken English," *Journal of the Acoustic Society of America,* vol. 35, pp. 892–904, 1963.

[4] J.L. Elman "Finding Structure in Time," *Cognitive Science,* vol. 14, pp. 179–212, 1990.

[5] T.D. Hanley, J.C. Snidecor, and R.L. Ringel "Some acoustic difference among languages," *Phonetica,* vol. 14, pp. 97–107, 1966.

[6] B.L. Kalman, S.C. Kwasny, and A. Abella "Decomposing Input Patterns to Facilitate Training," in *Proceedings of the World Congress on Neural Networks,* vol. III, Hillsdale, New Jersey: Lawrence Erlbaum Associates, 1993, pp. 503–506.

[7] S.C. Kwasny, B.L. Kalman, W. Wu, and A.M. Engebretson "Identifying Language from Speech: An Example of High-Level, Statistically-Based Feature Extraction," in *Proceedings of the 14th Annual Conference of the Cognitive Science Society,* Hillsdale, New Jersey: Lawrence Erlbaum Associates, 1992.

[8] H. Kucera and G.K. Monroe "A comparative quantitative phonology of Russian, Czech, and German," New York: American Elsevier, 1968.

[9] Y.K. Muthusamy and R.A. Cole "Automatic Segmentation and Identification of Ten Languages Using Telephone Speech," in *Proceedings of the 1992 International Conference on Spoken Language Processing,* vol. 2, 1992, pp. 1007–1010.

[10] Y.K. Muthusamy, R.A. Cole, and B.T. Oshika "The OGI Multi-Language Telephone Speech Corpus," in *Proceedings of the 1992 International Conference on Spoken Language Processing,* vol. 2, 1992, pp. 895–898.

[11] W. Wu, S.C. Kwasny, B.L. Kalman, and A.M. Engebretson "Identifying Language from Raw Speech: An Application of Recurrent Neural Networks," in *Proceedings of the 5th Midwest Artificial Intelligence and Cognitive Science Society Conference,* 1993, pp. 53–57.

CUBICORT: A Simulation of a Multi-column Model for 3D Image Analysis, Understanding & Compression for Digital TV, HDTV & Multimedia

P. LERAY *, Y. BURNOD **

* *Head of Advanced Research in Image Processing Laboratory, CCETT/TAV/CEP/ETS: Centre Commun d'Etude de Télédiffusion et Télécommunications - BP 59, 35510 Cesson-Sévigné, France.*
** *Head of Computational Neurosciences Group, Univ of PARIS VI*

Abstract

We describe here simulation elements and results of a new kind of 3D vision vimulator, for pre-processing in 3D object & movement analysis, using the biological concept of the cortical column paradigm in the primary visual area. The target simulator is primarily dedicated to ultra high image compression for *Telecommunication of TV images* (*MPEG4*), HDTV, and 3D TV, but can also be used for automatic modelling, digitizing, robotics or medical applications.

1 Introduction

Over the years, in depth study of 3D image analysis has been carried out using algorithmic procedures. Some interesting results are now being achieved, but the problem as a whole (I.E with natural, unknown 3D scenes) remains unsolved.

However the human visual system is able to analyse and extract 3D objects and movements. Several implementations have been proposed for artificial retinas. We present here simulation elements and results to map the properties of the primary visual area, which could be used generally for pre-processing, robotics, image analysis (i.e. for digitizing images), image synthesis (C.G.I), and for telecommunication & storage / retrieval problems such as image compression.

As neuroScience knowledge is constantly improving, it now seems to be possible to design a VLSI chip set which could perform in real time some of the main functions of the primary visual cortex.

2 Description

Such a system is currently simulated at the CCETT and is based on a model of cortical columns [2] and a 3D real time image synthesis machine [4]; [5].

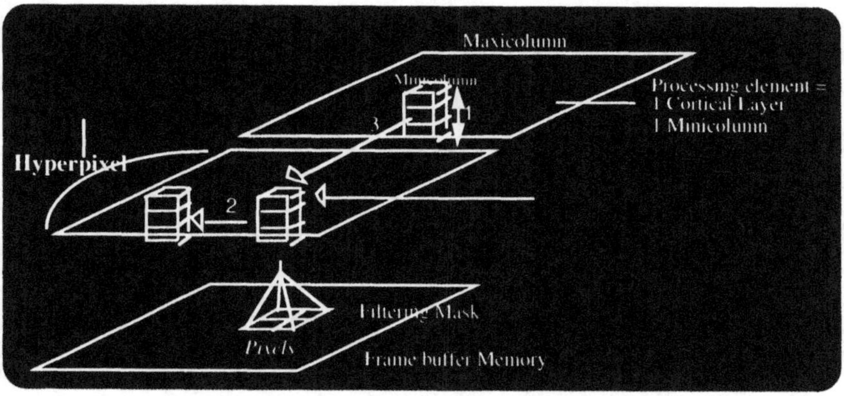

Figure 1: The Cortical Model: General Scheme

This machine consists of 4 layers defined by their primary index: 1: Pixels, 2: Zones (or hyperpolygons: which consists of a set of perceptually homogeneous zones of pixels.), 3: Objects and 3: Prototypes (or Labels). (See Figure 5).

2.1 The Pixel Level

Is a pre-processing, multi-scale pyramidal system which computes spatio-temporal filtering models of primary visual areas (V1 in particular) . In this paper, an extended concept of the conventional pixel is called Hyperpixel: In the brain, images received on the retina are transformed with elementary functions or filters in the cortical maps, which generate other information. Here the information extracted from a given retinal zone (receptive fields) is a hyperpixel; A set of hyperpixels is a hypermap; These elementary functions which are biologically performed in the visual cortex can be summarized as: i spatial functions: 2D gabor filter , (First and second order derivatives) and ii temporal functions (temporal derivatives or correlations between like features shifted in time and sopace). Each function is computed at 5 spatial resolutions, and 5 temporal resolutions. Spatial & temporal resolutions are directly linked and decrease simultaneously.

—Suppose we define a N*N pixel image on the input layer: 2 images at t & t+1 are stored, because high spatial resolution does not need a large number of successive images.

—At scale **s**, we will have $N^2/4^s$ hyperpixels stored for each image, and the resolution on the temporal axis will be $4^s * 33$ mS.

At this level **s**, we store 4^s images for temporal information, and $N^2/4^s*4^s = N^2$ hyperpixels. Each hyperpixel can be extracted according to different resolutions with the use of a special addressing mode in one memory cycle:

At each level, the effective address is shifted by 2 in both the X and Y direction. Each processor can access a block of hyperpixels and blur the less significant bits for the lowest resolutions. Global symmetry is achieved between the X, Y resolution, and time. Adjustable masks are stored at each level of the machine, and for each processor. A specific training process can be used in order to adapt each mask interactively according to the lower hyperpixels' activities. In this respect, masks can be modified and considered as synaptic weights. At this level, a superviser scans activities, and enables masks at the upper levels to be changed.

The first layer uses Gabor oriented filters (Product of a Gaussian by a sine) and Laplacian filters (Figure 2). The results are stored in a set of hyperpixels.

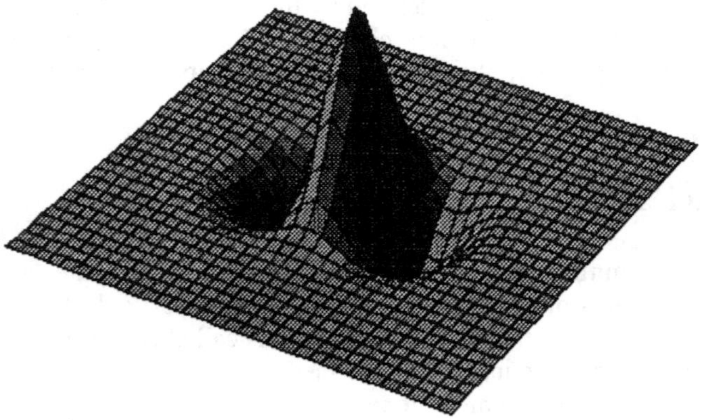

Figure 2: Directional Filters: First Level

From Gaussian and Laplacian filters we can at this level extract curvatures which are directly related to the Attention Points: These points correspond to highly informative zones of the picture.

Let: G = Gaussian L = Laplacien; $K = (d^2G/dx^2 - d^2G/dy^2)/2$,
$\text{Theta1} = 1/2\text{Arc}\cos\,(dG/dx)^2/\text{Gradient})$; $\text{Theta2} = 1/2\text{Arccos}\, d^2G/dxdy/R$;
The Curvature C is defined by:

$$C = (L+R*\cos(2*\text{Theta2} - 2*\text{Theta1} + PI/2))/\text{Gradient}$$

Attention points are defined by hyperpixels having a high degree of C activity (See Figure 3).

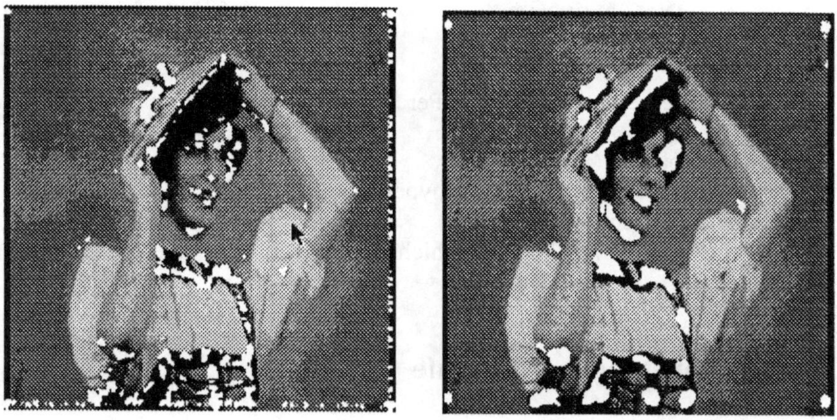

*Attention points with 16*16 filter* *Attention points with 32*32 filter*

Figure 3: Cortical columns: First level: neural filtering + attention points = informative zones hierarchy for coding

At this level perceptive grouping is also performed (See Figure 4), which gives a first estimation of homogeneous textures. Perceptive grouping can be modelled by an iterative process taking into account the successive modification of the model's filter column activity. At each image at time t, the activity Aj of the hyperpixel with index j can be expressed as follows:

$$A_j(t) = A_j(t-1) + \Sigma_{i=1,4}\,(\Phi(C,Ci)*A_{ji}(t-1))$$

Where Φ is a function of 2 columns. Development of activities Aj leads to each perceptually homogeneous zone being transformed in a uniform area.

2.2 The Upper Levels

The upper levels analyse images with unknown coefficients:

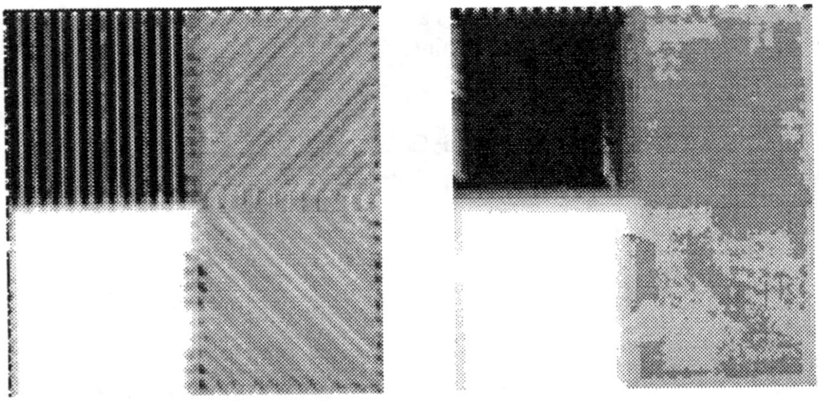

Figure 4: Perceptive Grouping: Perceptually Homogeneous Zones

- The hyperpolygon (zone) level,
- The object level, which is set of hyperpolygons. At this level, objects are viewpoint dependent.
- The prototype (or label) level which is a collection of objects viewpoint independent.

2.3 A Reconstruction Module

The reconstruction module draws 2D or 3D areas. At each level of the system, a reconstruction process is used; (feedback image synthesis). Reconstructed Image is compared with analysed image. The general scheme of a coding system is shown in Figure 5.
At each level the Cubicort structure is composed of:
- An input Hypermap for the input Image (feedforward process)
- An hypercolumn for the hyperpixel process, and directed by the primary index Basically the hypercolumn is built with a set of multiplier/accumulators, and a learning block which concatenates by a bi-linear function the different results obtained by other column Index and other layers;
- An output hyperMap for the output (reconstructed) Image (feedback process)
- The associated Hypercolumn.

Such a system can be used as a coding machine, a similar machine being used for the decoding process; the links between them being only the variations in activities of the different hyperpixels. Such a system can combine low and high level analysis and reconstruction.

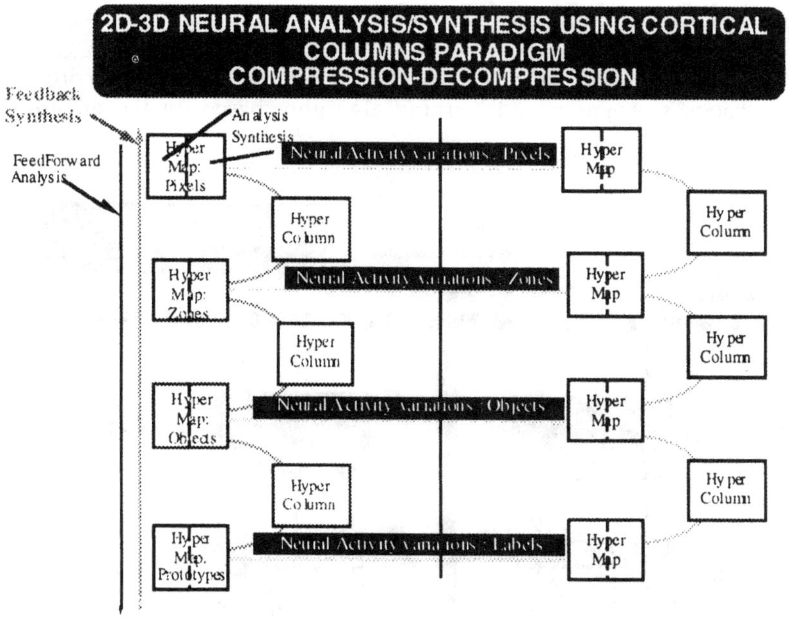

Figure 5: The Architecture.

However with this approach, we do not require full implementation of the 4 levels: For example, the filters used at the first level are roughly equivalent to the DCT, sub-band or wavelet coding algorithms.

Moreover, perceptive grouping and order of priority of informative zones defined by attention points already constitute a key element for coding enhancements: For these zones we can use standard coding schemes such as D.C.T. (MPEG), according to the speed of the region: the lower the speed, the smaller the compression ratio.

3 First Simulation Results:

3.1 Pre-Processing Simulation

The aim is to accelerate the filtering process by using accelerated multipliers/accumulators. We have now entirely simulated the first level, with results relating to perceptive grouping and attention points.

3.2 Level Two and Upper Level

At these levels, the synaptic coefficients are not known, but learned. The cortical paradigm proposes a generalized learning algorithm based on the diagram described in figure 6.

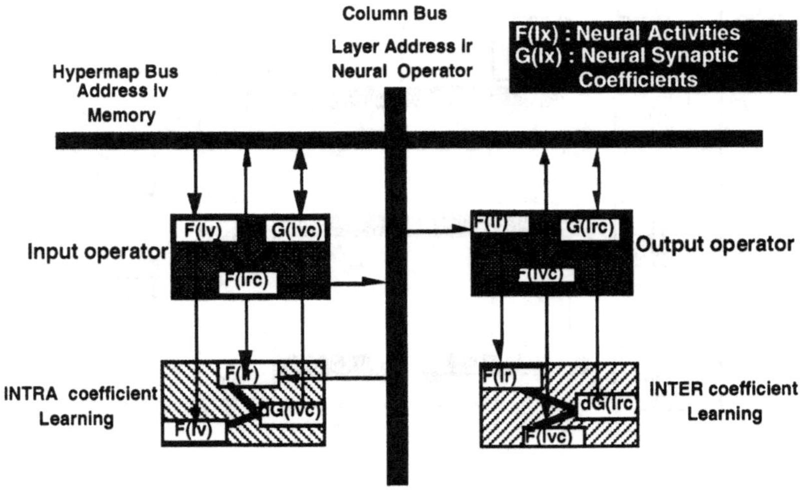

Figure 6: General Learning Architecture

This scheme is composed of two buses, the hypermap bus and the hypercolumn bus. Four columns are connected to these busses, and compute the sums of products. The different data in input and output are described in Figure 6. During the process, the hypercolumns continuously compute the different activities and synaptic weights which are stored in the hypermap. Learning and Image analysis are computed together.

- The result is that the system continuously adjusts its coefficients according to the input image. This system has been used and tested by [3].

3.3 Compression:

The complete coding scheme described in Figure 5 is now partly integrated: In particular, results of Level 1 will be implemented. Level 2 (Learning process will be integrated further).

4 Conclusion

These simulation results are a first step towards knowledge based multiscale vision analysis systems which could be used in ultra-high compression schemes as for future telecommunication standards within MPEG 4. Attention points and perceptive grouping are new key elements for giving relative informative importance to zones inside images. The cortical approach is moreover a generalization of DCT and transform coding schemes, with a fine tuning, of the image dependent filtering coefficients, and a cooperation of networks extracting various kinds of visual stable informations.
Since a symmetrical approach is used for analysis/synthesis, such a machine could also be used for object digitization/reconstruction in image synthesis techniques for digital TV and multimedia broadcasting.

Acknowledgements

We thank the CCETT and CNRS teams for their respective contributions. Particularly Messrs F. Guyot, F. Marchal, J.L. Anton, S. Guillaudeux, M. Klefstad and J. Nouvel.

References

[1] Alexandre, F., Guyot, F., Haton, J.P., Burnod, Y : "The cortical column, a new processing unit for multilayered network". *Neural Network*, Vol. 4, pp 15-25 1991.

[2] Burnod, Y. "An Adaptive neural network: the cerebral cortex." *Book: 400 pps. Masson; Prentice Hall* 1988.

[3] I. Otto, E.Guigon, P. Grandguillaume, L. Boutkhil & Y. Burnod, . "Direct and Indirect Cooperation between Temporal and Parietal Networks for Invariant Visual Recognition" . *Journal of Cognitive Neuroscience*, 4 (1) 35-57 (1992).

[4] Leray, P. "A 3D Syntheic Imagery Generator in real time" presented at the: *Image Generation/display Conference II* Af Human Resources Laboratory, Phoenix, AZ pp 78-89 1981.

[5] Leray P. "A 3D animation system Datastructures for Raster Graphics" *Data Structures for Raster Graphics Springer-Verlag* pp 165 - 171 1985 .

[6] Leray P. "Towards a Z-Buffer and Ray-Tracing Multimode System based on Parallel Architecture and VLSI chips", *Advances in Computer Graphics Hardware I Springer Verlag* pp 141 - 145, 1987

[7] Leray P. "Modélisation et architecture de machines de synthèse d'image pour la représentation et le rendu d'objets tri-dimensionnels sur écrans graphiques à balayage." *University of Paris II* , Doctorate 1990.

Self-organizing Finite State Vector Quantization for Image Coding

Hui Liu and David Y. Y. Yun

Department of Electrical Engineering
University of Hawaii at Manoa
2540 Dole St., Holmes Hall 491&2, Honolulu, Hawaii 96822 USA
Email: hliu or dyun@wiliki.eng.hawaii.edu

Abstract

This paper presents a new method in finite state vector quantization (FSVQ) for image coding by using self-organizing neural networks to automatically design the state codebook and next-state function without the need of matching the local statistics to the global statistics. It dynamically predicts the most probable codevectors for the current input as state codebook based on the distance information of codevectors used to encode the neighboring blocks. Our experiments show that, compared to memoryless VQ, the SNR improvement is up to 4.2 dB at the same rate, while the bit rate reduction is more than two times at the same SNR level. In addition, the proposed method is simple for hardware implementation.

1. Introduction

Vector quantization (VQ) techniques have been widely applied in source coding tasks such as speech and image coding [1]. An ordinary (memoryless) vector quantization exploits the high correlations between neighboring pixels within a block but ignores correlations between the neighboring blocks. To lower the bit rate further than is possible with a memoryless VQ, it is necessary to exploit the inter-block correlations with memory VQ, such as finite state vector quantization (FSVQ) [2-4]. The underlying problem of FSVQ is to maintain the quality (distortion) of a large codebook VQ while achieving the efficiency (rate) of a small codebook VQ. The major difficulties with FSVQ techniques are the design of state codebooks and the determination of the next-state function. Conventional FSVQ also requires huge amount of memory to store all of the state codebooks.

In this paper, we present a new design of FSVQ, called Feature Map VQ (FMVQ), for image coding by using self-organizing neural networks to automatically design the state codebook and next-state function without explicitly tracing the input statistics. Section 2 briefly discusses the FSVQ. FMVQ is presented in section 3, and the experimental results are presented in section 4.

2. Finite State Vector Quantization

The FSVQ encoder consists of a finite set of state codebooks where each state codebook belongs to a distinct state of the encoder. The state of a FSVQ encoder can be considered as a form of prediction of the next input vector based on the past encoded vectors or, equivalently, on the past reproductions available to the decoder. Clearly, choosing a good codebook is effectively equivalent to predicting the behavior of the next input vector given the current state.

FSVQ can be formally defined as below: An input vector X_n is encoded by applying the nearest neighbor search on the corresponding state codebook of the current state for the best representative codevector. The encoder can be considered as a mapping C, which converts an input vector into a channel index defined as

$$I_n = C(X_n, S_n) \tag{1}$$

where the encoder state S_n is defined using the next state function

$$S_n = F(I_{n-1}, S_{n-1}) \tag{2}$$

The next-state function $F()$ uses the previous state and the previous channel index to determine the current state of the encoder. Similarly, the decoder is a mapping D given by

$$\hat{X}_n = D(S_n, I_n) \tag{3}$$

where the reproduction vector (codevector) $\hat{X}_n$ is obtained by indexing the same state codebook as the encoder.

Generally, the design of a FSVQ is very complicated, involving three separate stages: (1) initial classifier design, (2) state space, next-state function and state codebook design, and (3) iterative state codebook improvement. Aravind and Gersho have first applied the FSVQ in image coding using the omniscient design [2]. The storage of state codebooks in their method is very large since there are many duplicated codevectors in those state codebooks. Kim [3] has developed a side-match rule to select codevectors from a super codebook (the collection of all codevectors in all the state codebooks) to form a state codebook in an attempt to force the encoder to optimize spatial contiguity across block borders without doing explicit edge detection. Nasrabadi and Feng [4] have proposed a dynamic finite state vector quantization which uses the directional condition block transition probability to reorder all the codevectors in the super codebook and to select the most probable codevectors as the state codebook. However, the reordering process is computationally expensive, resulting in the complexity of their algorithm approximately the same as a memoryless VQ with the super codebook size[4].

3. FSVQ using Self-Organizing Feature Map

The clustering property of competitive learning algorithms in neural networks appears to be well-suited for VQ codebook design [5-8]. However, little attention has been paid to apply the property of the self-organizing feature map in encoding/decoding phases. A topological feature map is essentially a mapping that preserves neighborhood relations in the input pattern space where the 2-D geometrical distance can be used to measure the similarity of patterns. Therefore, the location of the winning output unit conveys some information, with nearby outputs corresponding to nearby, i.e., similar, input patterns.

Figure 1 shows a codebook of 256 codevectors trained by Kohonen self-organizing learning algorithm [5] with a wrap-around neighborhood function. Those codevectors are arranged as a 16x16 grid. Clearly they are organized by similarity: similar codevectors (shapes and gray scales) are geometrically nearby. Note that codebooks trained by other algorithms do not have this property.

On the other hand, neighboring blocks in a natural image are often highly correlated (as a Markov source), thus, they are often quantized by nearby codevectors in the self-organizing codebook. Figure 2 shows the current block X with its four neighboring blocks A, B, C and D The codevectors used to encode the four blocks are called reference points of the input X. The neighborhoods centered at those reference points are π_i, $i = a,b,c,d$. By using the self-organizing super codebook, we can assume that the codevector that is the best representative of the current input block will be close to one of those reference points, i.e., the best match is most likely to be in the union of the four neighborhoods ($\Pi_X = \bigcup_i \pi_i$).

Figure 1. Super codebook trained by Self-organizing algorithm, where 256 codevectors are arranged as 16 by 16 grid, with each codevector as 4 by 4 block.

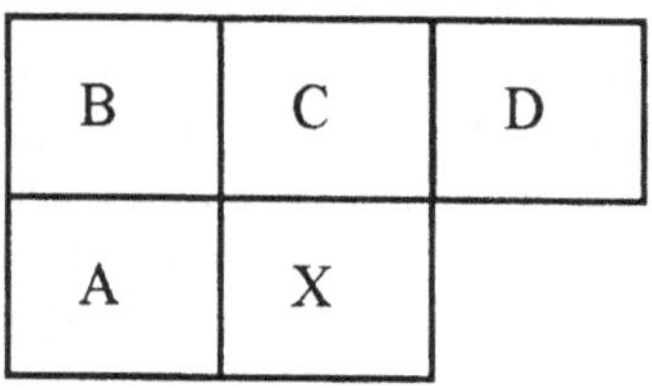

Figure 2. Neighboring Blocks

In terms of FSVQ, we may predict the most probable codevectors from the super codebook to form the state codebook based on their 2-D distances to those reference points. The state is determined by the positions of reference points and the corresponding state codebook is selected from the small portion of codevectors in Π_X. The decoder generates the same state codebook using these reference points.

This approach is totally different from traditional FSVQ methods since no statistics of input has to be calculated and no separate procedures are needed to design the state codebooks and next-state function. The self-organizing learning algorithm automatically captures the statistics of training data and converts the statistics into the 2-D topological relationships among the codevectors. This self-organizing property greatly simplifies the design of state codebook and next-state function.

Unlike other FSVQ design, the complexity of state codebook generation depends only on the state codebook size N_{state}. Thus, the encoding complexity of FMVQ is O(kN_{state}), where k is the codevector dimension. Compared to [4], where the complexity is O(kN_{sup}), FMVQ has significantly reduced the encoding complexity ($N_{sup} >> N_{state}$). In addition, FMVQ overcomes the problem of huge memory requirement for all state codebooks by avoiding duplicate codevectors.

When the neighboring blocks are not highly correlated, the best representative codevector may not appear in the state codebook. An adaptive FMVQ will first use the state codebook to encode the current input and if the distortion of the best match is higher than a preset threshold, it will use the super codebook for a better match. If the better match is found, than a flag with the index of the best match is sent to the receiver. Thus, the adaptive FMVQ can automatically switch between a FMVQ and a memoryless VQ.

4. Results and Discussion

Experiments on image coding have been performed using FMVQ. All images used for training and testing are 512x512 with 256 gray levels. The block size is 4 by 4. Ten images, including faces and outdoor scenes, are used to train the super codebook using Kohonen learning algorithm [5]. A wrap-around neighborhood function is used in to increase the codevector utilization. Three images, "Lena", "Peppers", and "Lake", that are outside of the training images, are used to test the performance. The sizes of super codebook vary from 64 to 4096, corresponding to 8x8 to 64x64 neurons in the competitive grid. The sizes of state codebook vary from 8 to 64.

Figure 3 shows the hit ratio (percentage of correct prediction) for the three test images. Clearly, the hit ratio depends upon the correlations in an image. The hit ratio increases with the state codebook size. For N_{state}=64, it is up to 94-98%.

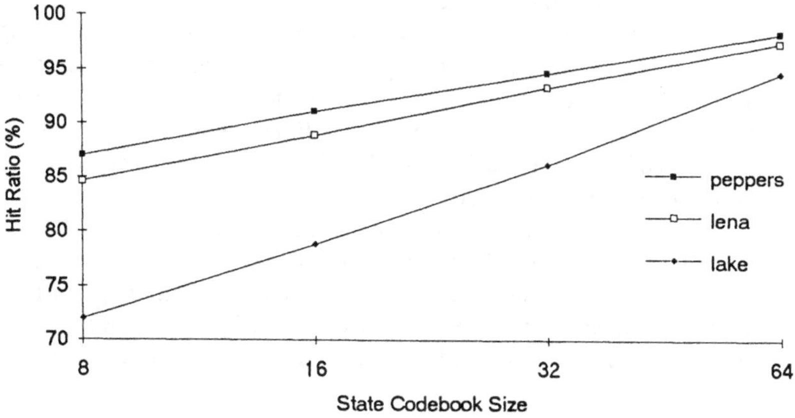

Figure 3. Hit Ratio vs. State Codebook Size

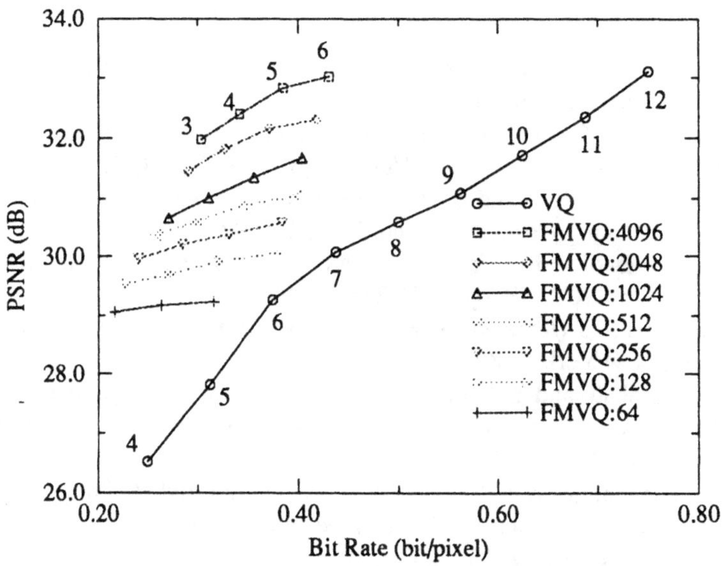

Figure 4. Rate SNR plot for test image "Lena"

Figure 4 shows the plot of bit rate against signal-to-noise ratio (SNR) for test image "Lena". The plot shows that, for a given super codebook, the PSNR increases at most 1 dB with 0.15 bpp cost as the state codebook size increases from 8 to 64. The SNR increases as the super codebook size increases. This suggests that large super codebook with small state codebook should be used to reduce the bit rate while increasing the image quality. Two observations are also found in Figure 4: First, for a given bit rate, the SNR improvement over a memoryless VQ is up to 4.2 dB. Second, at the same SNR level, the rate of FMVQ can be reduced more than half of the bit rate in memoryless VQ. The plot shows that at 32 dB, FMVQ with super codebook of 4096 and state codebook of 8 achieves 0.303 bpp rate and the corresponding VQ (by interpolation) takes 0.65 bpp.

Image "Peppers" and "Lake" have been used to study how the encoding performance varies from image to image. We have found that although the natures of these images are different, similar performance behaviors are obtained. These results prove that FMVQ is an effective and computationally efficient algorithm for image coding over a variety of images.

References

[1] A. Gersho and R. M. Gray, *Vector Quantization and Signal Compression*, Kluwer Academic Publishers, 1992

[2] A. Aravind, A. Gersho, "Image compression based on vector quantization with finite memory," *Opt. Eng.*, Vol. 26, pp. 570-580, July 1987.

[3] T. Kim, "Side match and overlap match vector quantizers for images," *IEEE Trans. Image Proc.*, vol.1, no. 2, April 1992

[4] N. M. Nasrabadi and Y. Feng, "A dynamic finite-state vector quantization scheme," *Proc. ICASSP*, pp. 2261-2264, 1990.

[5] T. Kohonen, *Self-organization and Associative Memory, 2nd ed.*, Springer, New York, 1988

[6] N. M. Nasrabadi and Y. Feng, "Vector quantization of images based upon the Kohonen self-organizing feature maps," *Proc. IEEE IJCNN*, Vol. 1, pp. 101-108, 1988.

[7] A. K. Krishnamurthy, A. C. Ahalt, D. E. Melton and P. Chen, "Neural networks for vector quantization of speech and images," *IEEE J. Select. Areas Commun.*, Vol. 8, No. 8, Oct. 1990.

[8] H. Liu and D. Y. Y. Yun, "Competitive learning algorithms for image coding,", *Proc. of SPIE on Applications of Artificial Neural Networks III*, SPIE Vol. 1709, Orlando, FL, April. 1992

A Learning Model for Adaptive Network Routing[†]

Tony R. Martinez and George L. Rudolph

Department of Computer Science
Brigham Young University, Provo, Utah 84602
martinez@cs.byu.edu
george@axon.cs.byu.edu

Abstract

Increasing size, complexity, and dynamics of networks require adaptive routing mechanisms. This paper proposes initial concepts towards a learning and generalization mechanism to support adaptive real-time routing. An ASOCS learning model is employed as the basic adaptive router. Generalization of routing is based not only on source/destination address, but also on such factors as packet size, priority, privacy, network congestion, etc. Mechanisms involving continual adaptation based on feedback are presented. Extensions to conventional addressing which can support learning and generalization are proposed.

1. Introduction

As networks increase in size and routing complexity, new types of mechanisms are required to support routing decisions. In particular, network routers must accommodate and adapt to rapid changes in network topology and usage.

This paper proposes a learning and generalization scheme to support adaptive real-time routing in computer networks. Traditional routing algorithms are based on tabular storage of destination/gateway tuples which decide where a packet should be forwarded upon arrival at a gateway. This method suffices when this tabular information is available and up to date. When not available,

[†]This research was supported in part by a grant from Novell Corporation

mechanisms such as default gateways, random forwarding, or message discarding are employed. We propose that generalization schemes may be used to forward packets in a more efficient and accurate manner when confronted with insufficient and inaccurate routing information.

2. Approach

Most current algorithms base routing decisions primarily on the destination address. Though destination address is obviously crucial, other information may be important in making a proper routing decision. For a packet, this information includes packet size, priority, preferences (latency, throughput, reliability, etc.), privacy classification, etc. For the network such information as congestion, current failures, load, etc., may be pertinent. We call these *packet features* and *network features* respectively. We refer to address, packet features and network features collectively as *routing features* in this paper. Ad hoc algorithms require unique handling of these different features. A more general learning scheme treats all of these uniformly, with weighting of importance taking place during learning, thus both simplifying and expanding the approach to routing.

McDonald [3] proposes an Adaptive Self-Organizing Concurrent System (ASOCS) model as an adaptive network router. ASOCS [4,5,6] are a class of connectionist learning models which seek computational efficiency as highly parallel and adaptive connectionist logic systems that map discrete input values to output values. An ASOCS can infer routing decisions from routing rules that have been expressed as discrete input variables. This paper seeks to extend this approach through more robust generalization techniques. Below, the adaptive mechanism used for learning routing information and doing actual forwarding is simply referred to as the *router*.

The approach proposed in [3] assumes a version of the learning bridge algorithm. A router observes the source address and routing features of each incoming packet and remembers on which interface it arrived. When the address and other features are later observed as a destination, then the router forwards the packet on the interface on which the address was earlier observed as a source address.

As new nodes are added to a network (or even with a network initialized with no routing information), these nodes can send out a *beacon* packet which is forwarded along all routing lines except that on which the beacon was received. This process "primes" the network with initial routing information which is learned as the network evolves. By adding distance information (number of hops, time lapse, etc.) to the beacon packet, an accurate assessment of efficient routing is captured. However, accuracy of information may still be faulty due to the following.

- The beacon process above must be limited because of excessive bandwidth requirements.
- Conditions at the time of the initial beacon were not representative of

average case.

- A constantly changing network causes initial information to be inaccurate.
- The initial beacon is based only on source address and not on other routing features.

These and other factors cause the necessity of ongoing learning and adaptation to the network environment. The model above provides a reasonable initial state from which continual adaptation and refinement can occur. This adaptation requires some source of feedback to the router to allow update of its current state. Some potential approaches include:

- Feedback Protocol: This protocol would cause (optionally) the router which received the packet to send information back to the source address regarding the "cost" of receiving the packet. This cost is then used to update and improve the initial sending router. Note that cost in this context may involve not only time to arrival, but metrics relating to errors, privacy, reliability, etc.
- Piggy-Back Metrics: A feedback protocol will add to network traffic. Another option is to maintain the current cost metric with the packet. Upon arrival at a router, if this metric is excessive are not in line with a *predicted* cost stored at the receiving router, only then need this information be sent back to the source routers.
- Inquiry Packets: Discovery of improved routing can be initiated by a router through the sending of *inquiry* packets which are a more selective version of beacon packets. In particular, a router with poor predictive cost may seek improved routes to commonly occurring destinations. These can be initiated at times of lower network load so as to seek constant improvement without an adverse effect on network traffic.

3. Learning and Generalization

There are a number of learning and generalization models which could be used to support generalization of network routing. In the case where a gateway has seen an incoming source address and later needs to send to that address as a destination, the mechanisms of backward learning overviewed above could suffice. However, there are times when a source needs to send to an address from which it has not yet received a message and thus no information regarding that address will be found in the gateway's current routing tables. This lack could be due to newness of the destination site, error at the routing table, etc.

In this case the address given by the initial sender could not just be the standard symbolic name (i.e. jdoe@cs.univ.edu), since the name server is not yet aware of this address. The numeric name could be given and although the gateway has no record of this yet, generalization could be done to make a best guess at routing, rather than the standard return of host unknown. However, since numeric addresses are typically given randomly to new hosts, no interesting generalization can be made from the address number itself. For example, a learning model could note that addresses 123, 129 and 140 are best delivered to output line 1 and 124, 160, and 162 to output line 2. If a new address, 144, is the destination and it has not been previously seen, the learning model could only make an arbitrary guess as to the best output line. This is because the input feature is a random "address" which has no regularity or critical sub-features which can be learned to discriminate or join itself with/from other addresses. As mentioned above, other routing features can be used to aid in generalization, but without at least some candidate output lines based on address the other information is moot.

For any learning system to actually be able to generalize on inputs not seen before, the input must include features which have within them some kind of regularity which the learning model can extract. We propose that to support generalization on unknown addresses the address itself must contain some extractable regularities. For example, a uniquely identifying address (which could be a random set of bits) could be augmented to also contain geographical information, such as the approximate latitude and longitude of the address site. (Similar extensions having different motivation and advantages are proposed elsewhere [1,2]). In this case, there is extractable regularity to be learned from addresses both incoming and already available in the routing table. When a new address is the destination, one possible learning bias is that the packet should be sent on an output line which has been successful for addresses with a similar geographic location as the new one. At this point, deciding between candidates can be enhanced using other routing features which are a part of the packet itself.

Other regular features could be part of the basic address. One which is already part of current internet addresses is the organization type (i.e. edu, com, gov). This type of information can be used to aid in learning what types of lines (security, speed, size, etc.) are typically best suited to communication involving these classes of users.

While too much addition of regularity would lead to cumbersome addresses, a judicious extension to the basic address mechanism could allow generalization on routing to unknown locations with reasonable cost and accuracy.

4. Minimization

While not supporting full generalization as in the previous section, the learning bridge algorithm can lead to more concise routing information for addresses actually received. Hierarchical IP addresses (i.e. address prefixes, geographical information, or an equivalent labeling system) could provide hierarchical abstraction through the process of minimization. A pair of prefixes of length N can be minimized to a single prefix of length N-1 if the prefixes differ in only one bit (e.g. the prefixes 1010 and 1011 can be summarized to the prefix 101*, where * is a don't care variable). This can be repeatedly applied so that a set of longer prefixes can be combined into a single, much shorter prefix. (Note that this same philosophy can be used with the geographical extension of section 2 to allow the representation of a large number of routing tuples into a single rule based on geography or other feature.) This includes prefixes for both route "table" entries (internal minimization), and route update information sent across the network (external minimization). Minimization thus helps reduce growth and size of routing tables, as well as route information traffic in the network. If eight sites are attached to the same network regional router, and their prefixes can be minimized, then the regional router can store, and advertise, its reachability to all eight sites with a single prefix. As a result, the number of routing entries and routing information traffic will scale less than linearly with respect to the number of networks.

5. Conclusion and Current Work

This paper presents initial approaches which can support routing generalization for novel network addresses. While these ideas are still in an early stage, current work seeks to demonstrate potential through network routing simulation and generalization studies

References

[1] Ford, Peter S., Yakov Rekhter, and Hans-Werner Braun, Improving the Routing and Addressing of IP, *IEEE Network*, **Vol. 7**, #3, pp. 10-15, May 1993.

[2] Katz, Dave, and Peter S. Ford, TUBA: Replacing IP with CLNP, *IEEE Network*, **Vol. 7**, #3, pp. 38-47, May 1993.

[3] Mcdonald, K., T. R. Martinez, and D. M. Campbell, A Connectionist Method for Adaptive Real-Time Network Routing, *Proceedings of the 4th International Symposium on Artificial Intelligence*, pp. 371-377, 1991.

[4] Martinez, T. R. and Vidal, J.J., Adaptive Parallel Logic Networks, *Journal of Parallel and Distributed Computing*, **vol. 5**, 26-58, 1988.

[5] Martinez, T. R. and Campbell, D. M., A Self-Adjusting Dynamic Logic Module, *Journal of Parallel and Distributed Computing*, **vol. 11**, no. 4, pp. 303-313, 1991.

[6] Martinez, T. R. and Campbell, D. M., A Self-Organizing Binary Decision Tree for Incrementally Defined Rule Based Systems, *IEEE Transactions on System, Man, and Cybernetics*, **vol. 21**, 1991.

Multilayer Perceptron based Equalizers applied to Nonlinear Channels

Michael Meyer and Gerd Pfeiffer

Department of Communications Engineering
University of Paderborn
33095 Paderborn, Germany
`nmeye1@pbhrzt.uni-paderborn.de`

Abstract

The application of multilayer perceptrons to equalize time-invariant nonlinear channels with intersymbol interference is described. The multilayer perceptron is used in a decision feedback structure according to a conventional decision feedback equalizer. Simulation results show that the bit error rate performance of the neural network is significantly superior compared with the decision feedback equalizer if nonlinearities influence the transmission. The performance improvement depends on type and strength of the nonlinearity.

1 Introduction

Decision feedback equalizers (DFEs) are often applied in digital communications systems to reduce the influence of channels producing intersymbol interference. Since channel equalization can be seen as a classification problem, an approach of using neural networks for this application was considered. In earlier studies the performance of multilayer perceptron (MLP) based DFEs was investigated for linear channels [1].

This paper describes the use of this type of neural network on extended equalization problems. Here, signals disturbed not only by intersymbol interference but also by nonlinearities in the transmission channel are considered. In particular, different types of nonlinearities are investigated to give an overview on their influence on data transmission.

In section 2 the assumed data transmission system is described. The subsequent section presents the MLP and its application on channel equalization. The fourth part gives simulation results and a bit error rate performance comparison between the conventional decision feedback structure and its neural network counterpart.

2 Data Transmission System

Figure 1 shows the block diagram of the used baseband data transmission system. The information source provides a binary information sequence $\{q_k\}$. By using quadrature phase shift keying (QPSK) modulation two consecutive binaries form the complex signal $\{s_l\}$. The index l indicates the symbol interval T, whereas k denotes the bit interval $T_b = T/2$ of the source. The signal $\{s_l\}$ is transmitted over an equivalent discrete-time channel with the impulse response $\{h_l\}$ [2].

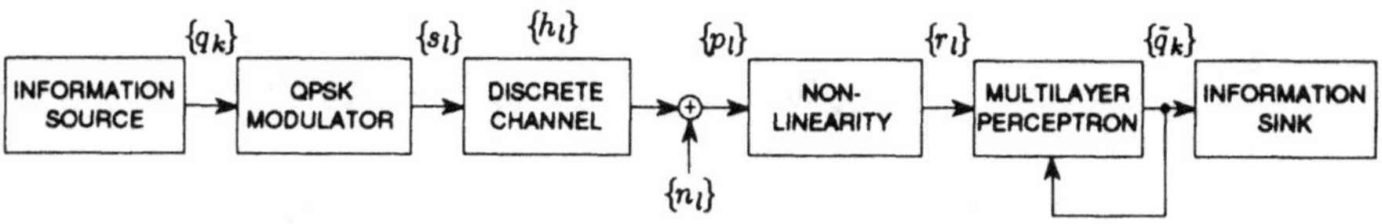

Figure 1: Structure of a baseband data transmission system

Additionally, the transmitted signal is corrupted by complex additive white Gaussian noise $\{n_l\}$ with zero mean and two-sided spectral density $N_0/2$. Therefore the signal $\{p_l\}$ can be written as

$$p_l = \sum_{i=0}^{L} s_{l-i} h_i + n_l, \tag{1}$$

where L gives the number of symbols which are affected by intersymbol interference. So far, the transmission system is considered to be linear. Then a nonlinearity distorts the signal. This paper investigates as a examples two types of nonlinear functions: $G_1(g\xi) = \tanh(g\xi)$ and $G_2(g\xi) = g\xi^3$, where g is an additonal parameter to vary the strength of the nonlinearity.

The received signal $\{r_l\}$ is fed into the MLP, which tends to remove the disturbances and demodulates the signal. After a decision device the detected information sequence $\{\tilde{q}_k\}$ is compared with the sequence $\{q_k\}$ to determine the bit error rate.

3 Multilayer Perceptron based Decision Feedback Equalizer

The MLP is a well-known type of neural network [3, 4]. Its main characteristics are feed-forward structure, arrangement of multiple neurons in layers and backpropagation training algorithm [5]. Figure 2 presents the structure of a MLP with M layers.

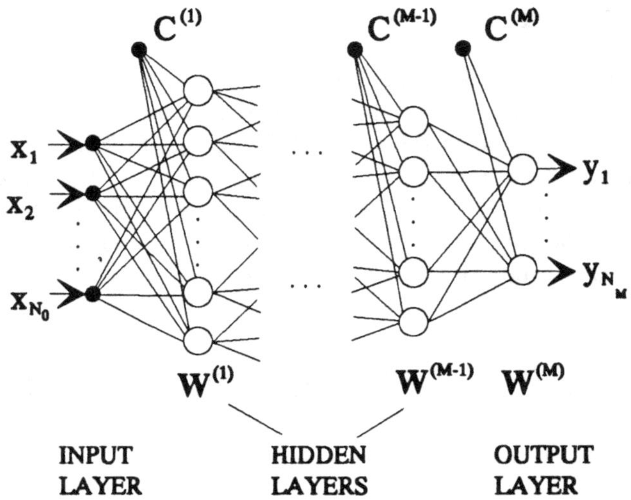

Figure 2: Architecture of a MLP with M layers

The structure and function of MLPs is assumed to be known. Hence, this paper concentrates on the application of the neural network. The input layer receives the input vector $\mathbf{x} = [x_1, x_2, \ldots, x_{N_0}]$ and the network estimates the output vector $\mathbf{y} = [y_1, y_2, \ldots, y_{N_M}]$. To enable the MLP to perform its task, the desired mapping has to be specified through a training phase. Therefore the backpropagation algorithm is applied [5]. The desired symbols are compared with the outputs of the neural network, resulting in an error signal used to adapt the weights of the MLP. After learning input data can be classified by the network.

As described in section 2, the task of the MLP is to remove intersymbol interference and distortions due to nonlinearities. For that purpose the integration of the neural network into the transmission system (Figure 1) is depicted in Figure 3.

The received complex symbols $r_l = r_{l,I} + jr_{l,Q}$ are divided in their inphase and quadrature component $r_{l,I}$ and $r_{l,Q}$. In analogy to conventional

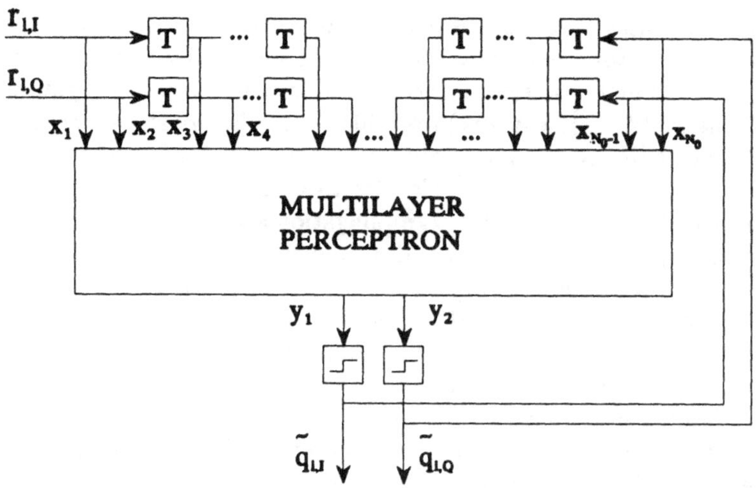

Figure 3: Structure of an MLP based DFE

DFEs [6], the structure consists of a feedforward part and a feedback part. The inputs to the feedforward tapped delay line are received complex symbols, and the inputs to the feedback part are hard-limited complex output symbols. Since the MLP is used to equalize QPSK-modulated symbols, the network provides two output values y_1 and y_2 for each symbol interval.

4 Simulation Results

Computer simulations of the data transmission system described in section 2 were performed to demonstrate the performance of MLPs. Here, the applied discrete-time channel is given by the z-transform [6]

$$H(z) = 0.407 + 0.815z^{-1} + 0.407z^{-2}. \tag{2}$$

A MLP based DFE is denoted by (a, b)-MLP, assuming that the feedforward part receives a and the feedback part b complex symbols, or analogous (a, b)-DFE in the case of conventional equalizers. A (3,2)-MLP with two layers was used to equalize the channel given in eq. (2) and the results were compared to those obtained in simulations of a (3,2)-DFE. Further investigations, not presented here, have shown that a MLP with two layers offers the same performance as a MLP with three layers for this application [7].

All simulations were performed with the same configuration and the same parameters. According to the (3,2)-structure, the MLP consisted of 10 neurons in the input layer. Additionally, 8 neurons in the hidden layer and 2 neurons in the output layer were used. The applied activation function was $\tanh(\cdot)$. This function is symmetrical to the origin and maps input values into the interval (-1,1), which is preferable for this application, because the signals are distributed symmetrically, too. Both, the MLP based DFE and the DFE were trained with 10^5 binaries to ensure optimal adaptation to the channel. According to the backpropagation algorithm described in [3], a learning gain factor $\eta = 0.005$ and a momentum factor $\alpha = 0.5$ for the neural network and the LMS-algorithm [6] with step size $\Delta = 0.005$ for the conventional equalizer were used. After training, the bit error rate P_b was measured. The given results are the average of 100 individual runs of 10^5 binaries, each started with new random weigths and random sequences.

A performance comparison between (3,2)-MLP based DFEs and conventional DFEs on equalizing linear channels leads to the conclusion that they provide nearly the same bit error rates as depicted in Figure 4.

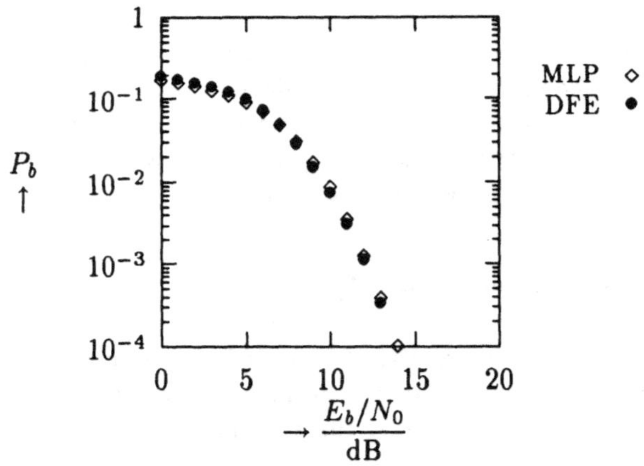

Figure 4: Performance of (3,2)–DFE and (3,2)–MLP, linear system

If the transmission system contains nonlinearities, the performances of both equalizers show significant differences. Figure 5 depicts the functions $G_1(g\xi) = \tanh(g\xi)$ and $G_2(g\xi) = g\xi^3$ for different values of g used in all simulations. Inphase and quadrature components of the signal $\{p_l\}$ are separately used as arguments of the nonlinear functions.

The curves indicate distinctly how the strength of the nonlinearities is influenced by the parameter g, especially for $G_1(g\xi)$. Each function has its own special characteristic. $G_1(g\xi)$ maps arguments to the restricted interval (-1,1), whereas $G_2(g\xi)$ attenuates small values and amplifies higher values.

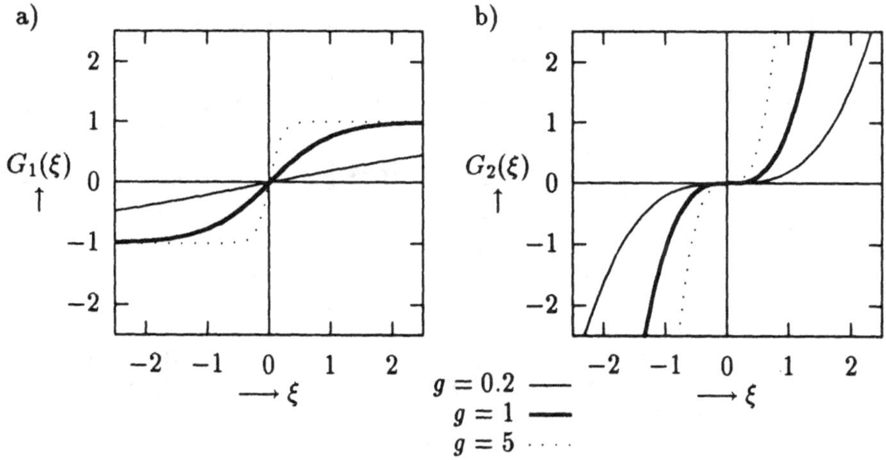

Figure 5: a) $G_1(\xi) = \tanh(g\xi)$, b) $G_2(\xi) = g\xi^3$

Figure 6 gives the simulated bit error rates P_b for the MLP based equalizer and the conventional DFE for various signal-to-noise ratios and three different gain factors g.

The curves for the (3,2)-MLP (Figure 6a) make it evident that the bit error rate performance for $g = 0.2$ is similar to the performance for the linear system and increases for stronger nonlinearities. Although the same statement is valid for the DFE, the results for the (3,2)-MLP show a significant superiority of the MLP for stronger nonlinearities ($g = 1, g = 5$). This is due to the ability of MLPs to build nonlinear decision regions, and therefore a better adaptation to the desired mapping of input values into output values can be obtained.

In Figure 7a the results for the MLP are presented for $G_2(\xi) = g\xi^3$. The neural network based equalizer is able to classify the received signal although there exists a significant deterioration against the linear system. Compared with Figure 4, the performance loss for $g = 1$ and $P_b = 10^{-3}$ is about 4 dB. The gain factors $g = 0.2$ and $g = 5$ lead to worse bit error rates.

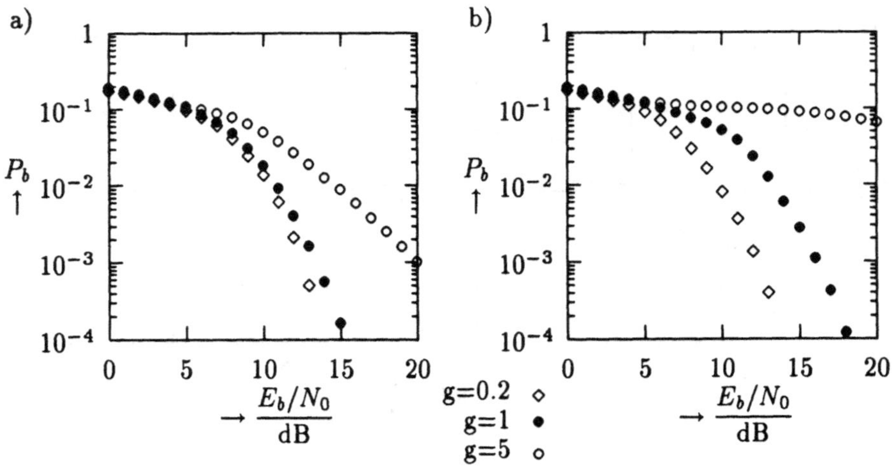

Figure 6: Performance for $G_1(\xi) = \tanh(g\xi)$: a) (3,2)–MLP, b) (3,2)–DFE

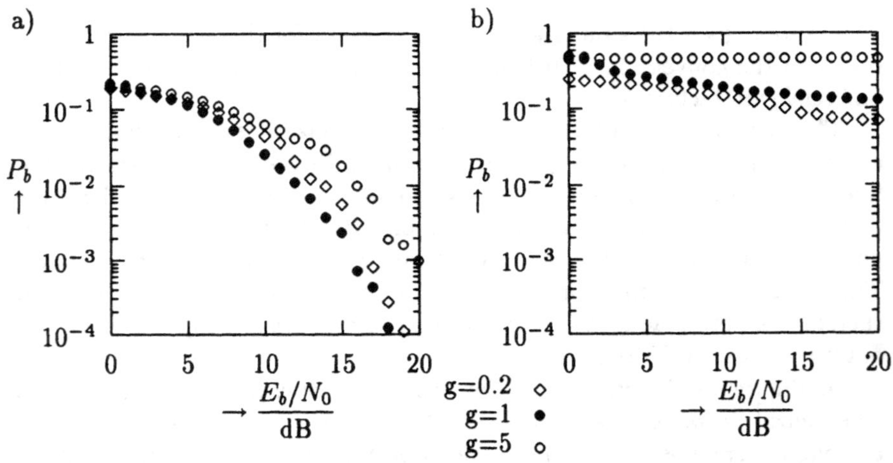

Figure 7: Performance for $G_2(\xi) = g\xi^3$: a) (3,2)–MLP, b) (3,2)–DFE

Figure 7b illustrates the performance of the conventional DFE for the same simulation parameters. Obviously, this structure is not suitable for equalizing this type of disturbances. For example, the bit error rate is $P_b \approx 0.5$ for $g = 5$ independent of the signal-to-noise ratio in the range considered here.

Further simulations with other types of nonlinearities were performed, which are not presented here.

5 Conclusions

This paper has presented the application of MLP based DFEs and conventional DFEs on equalizing nonlinear channels with intersymbol interference. Two types of nonlinear functions were considered.

The results on equalizing linear channels (Figure 4) show that there is no significant difference in bit error rate performance between the two equalizer structures. If the transmission system contains nonlinearities the MLP offers a great superiority compared with the conventional structure. Only for quasi-linear functions is the performance equivalent. However, it has to be stated that the performance improvement of the MLP based equalizer is achieved with a much greater computational expense.

References

[1] S. Siu, G.J. Gibson, C.F.N. Cowan, Decision Feedback Equalisation using Neural Network Structures and Performance Comparison with Standard Architecture, *IEE Proceedings*, Vol. 137, No. 4, pp. 221–225, August 1990.

[2] G.D. Forney, Maximum-Likelihood Sequence Estimation of Digital Sequences in the Presence of Intersymbol Interference, *IEEE Transactions on Information Theory*, Vol. IT–18, No. 5, pp. 363–378, May 1972.

[3] R.P. Lippmann, An Introduction to Computing with Neural Nets, *IEEE ASSP Magazine*, Vol. 4, pp. 4–22, April 1987.

[4] P.D. Wassermann, *Neural Computing – Theory and Practice*, Van Nostrand Reinhold, New York, 1989.

[5] D.E. Rumelhart, G.E. Hinton, R.J. Williams, Learning Representations by Back-Propagating Errors, *Nature*, Vol. 323, pp. 533–536, October 1986.

[6] J.G. Proakis, *Digital Communications*, McGraw-Hill, 2. Ed., 1989.

[7] M. Meyer, Verwendbarkeit neuronaler Netzwerke zur Entzerrung und Demodulation digitaler Signale, Diplomarbeit 7/91, Universität–GH–Paderborn, 1991.

Quo Vadis – A Framework for Adaptive Routing in Very Large Communication Networks

Armin R. Mikler, Johnny S.K. Wong, Vasant G. Honavar

Department of Computer Science
Iowa State University
Ames, Iowa 50011, USA
mikler@cs.iastate.edu
wong@cs.iastate.edu
honavar@cs.iastate.edu

Abstract

Quo Vadis is a heuristic approach to routing in a modern communication environment consisting of a very large number of nodes. It is aimed at the elimination of expensive and unwieldy routing tables thereby reducing avoidable message overhead. Routing decisions are based on parameterized knowledge representation and heuristics which encode a limited view of the network as seen from each of the nodes. This forms the basis for future work on learning mechanisms that can be used to determine suitable parameter settings thereby making the network adaptive and hence responsive to complex network dynamics.

1 Introduction

Modern communication networks can be expected to contain hundreds or even thousands of network nodes to which thousands of user nodes are connected. In such a network, the issues of routing, flow control, reliability, and traffic management need to be re-examined as the degree of uncertainty of information upon which decisions are based increases significantly with increasing network size. Quo Vadis utilizes insights from a number of hitherto disparate areas, namely, communication networks, artificial intelligence, and adaptive dynamic systems in the form of a novel approach to answering such questions.

Conventional approaches to these issues involve the transmission and storage of network information about every node in the network. However, due to the size and the dynamics of large high speed networks, resource requirements such as storage and bandwidth, become prohibitively large. Furthermore, such approaches entail significant overhead which reduces the overall network utilization. These problems had been recognized in the mid 1970s, initiating the development of strategies to reduce resource requirements for traffic management. The resulting approaches involved subnetworking at both the physical level as well as the logical level. Hierarchical Routing [2] as well as Landmark Routing [3]. are strategies that involve the hierarchical structuring of a large network into logical units; Nodes within a logical unit keep *precise* state information about a limited set of nodes within the unit. While hierarchical structuring does indeed reduce the amount of information that has to be stored in nodes, it gives rise to a variety of new problems [4]. Limited state information has been shown to have the adverse effect of increasing the average path length between source and destination nodes [3]. The fact that an expanding communication environment requires periodical restructuring also presents a major drawback.

Quo Vadis represents a novel approach to routing in large communication networks motivated by the need to work with imprecise and uncertain information. Routing decisions in Quo Vadis are made by employing appropriate heuristics suggested by the nature of the task. For example, the farther a message is from its final destination, the greater is the likelihood of having different paths of comparable quality (with respect to the particular metric used) available for routing a message towards its destination [1]. This leads to one of the key ideas behind Quo Vadis, namely, to utilize local state information in conjunction with a less precise *view*, a summary, of the state of the entire network. This view, based upon measurements obtained during the course of normal network operation, is limited in its precision yet adequate to make appropriate routing decisions. Network information to be transmitted between and stored within nodes can thus be reduced significantly.

2 Quo Vadis

Quo Vadis employs the principles of queueing theory as each node is modeled as a M/M/1 queue. Let $l_{i,j}$ denote the link connecting nodes n_i and n_j with capacity $c_{i,j}$ and current flow $x_{i,j}$. Further, let L denote the set of all links $l_{i,j}$ in the network. Node n_i's average service rate μ_i and arrival rate λ_i are given by:

$$\mu_i = \sum_{\forall l_{i,j} \in L} c_{i,j} \qquad and \qquad \lambda_i = \sum_{\forall l_{j,i} \in L} x_{j,i} \tag{1}$$

A node's utilization or load is represented by $\rho_i = \lambda_i / \mu_i$.

Node n_i will compute its view $\mathcal{V}_i$ of the network's load conditions as $\mathcal{V}_i = \mathcal{F}(\mathcal{V}_j, \rho_j)$. Here, $\mathcal{F}()$ appropriately weights with $\mathcal{V}_j$ and ρ_j). For Quo Vadis, we have restricted the granularity of the view $\mathcal{V}_i$ computed by $\mathcal{F}()$ to a north-south-east-west view.

The size of views (knowledge bases) in nodes is constant and depends solely on the number of neighbors. It will not grow with increasing network size. Let T be the total number of nodes in the network and c the average number of neighbors, then memory needed for conventional views (i.e., routing tables) is $O(T)$ while our approach yields a view-size of $O(c)$. For large networks, $c \ll T$, thus a heuristic approach would yield significant savings in storage and processing time.

2.1 Knowledge Acquisition and Representation

In routing mechanisms based on routing tables or topology information the very fact of the existence of a node is represented by an entry in the global view. Quo Vadis does not have such an inherent way to do so. This strongly impacts the addressing scheme used in the communication network. Nodes cannot simply be identified by a generic name; a node needs to be associated with a unique location. To do so, a grid structure has been superimposed upon the communication environment such that nodes can be addressed by their coordinates (i.e., X-coordinate and Y-coordinate). This, however, *does not* force links between nodes to follow the x-y-direction. Network nodes are thus referred to by using their coordinates as a node-id.

A *good* view computation strategy must yield a sufficiently accurate representation of network load information. Clearly, this can only be evaluated with respect to a particular network topology, network dynamic, and a specific routing algorithm that utilizes network information. As described earlier, a node n_i utilizes local measurements ρ_j as well as view information received from neighbors $\mathcal{V}_j$ to compute its view $\mathcal{V}_i$. This view can further be decomposed into $\mathcal{V}_j^n$, $\mathcal{V}_j^s$,$\mathcal{V}_j^e$, and $\mathcal{V}_j^w$, representing n_j's north, south, east, and west view, respectively. Within the view computation phase, a *current* view for every direction is calculated. Information used to compute the directional components of n_i's view is weighted according to the relative location of the neighbor n_j with respect to the view component to be computed. This weight is defined as the directional gain $\mathcal{G}_{i,j}^d$, where d represents the direction of the components (i.e., north, south, east, west).

Let n_i be the current node, and let n_j be any neighbor of n_i, and m be the number of all such neighbors. Let (X_i, Y_i) and (X_j, Y_j) be the coordinates of n_i and n_j respectively, and let $dist_{i,j}$ denote the Euclidean Distance between n_i and n_j. The directional gain, $\mathcal{G}^s_{i,j}$ towards a destination south of n_i through n_j is then computed as:

$$\mathcal{G}^s_{i,j} = \begin{cases} 1 + \frac{(Y_i - Y_j)}{dist_{i,j}} & \text{if } Y_i \geq Y_j \\ 0 & \text{otherwise} \end{cases} \tag{2}$$

The directional gains for the north, west, and east components are computed similarly. For normalization, we define

$$\mathcal{G}^d_i = \sum_{\forall n_j} \mathcal{G}^d_{i,j} \tag{3}$$

With d representing the current direction, any component of node n_i's view,$\mathcal{V}^d_i$, can then be computed as

$$\mathcal{V}^d_i = \sum_{\forall n_j} \frac{\mathcal{G}^d_{i,j}}{\mathcal{G}^d_i} * (\alpha * \rho_j + (1 - \alpha) * \mathcal{V}^d_j) \tag{4}$$

where α determines the decay of significance of an event in node n_j along a particular route. In the current form, the view computation weights directional network information according to relative position of its source. Equation (4) represents only one of many possible forms of view computation. Other forms will have to be investigated through simulation of a variety of different network environments.

After a current view has been computed for each node n_i in the simulated network environment, these views, together with local measurements, need to be distributed to neighbor nodes.

2.2 Making Routing Decisions

A node n_i makes a routing decision by selecting one of its neighbors to be the *next node* on the route to be established. Clearly, the choice of neighbor must depend on its promise to propagate information towards its destination along a path that is optimal with a certain evaluation criterion. This criterion is generally referred to a routing metric.

The routing mechanism in n_i must therefore compute a *utility* or promise for each of its neighbors n_j with respect to a particular destination node n_d. The neighbor with the best utility will then be selected as the next node along the route towards n_d.

The Quo Vadis routing algorithm evaluates the utility of neighbor nodes n_j by taking their local measurements ρ_j as well as their view $\mathcal{V}_j$ into consideration. The directional components of $\mathcal{V}_j$ are being weighted with respect to the relative positions of n_j and n_d, resulting in a single view value v_j [5]. The load utility of node n_j can thus be computed as:

$$LU_j = \gamma * \rho_j + (1 - \gamma) * v_j \tag{5}$$

where $0 \leq \gamma \leq 1$. The tuning parameter γ determines whether the evaluation is biased towards a local optimization when $\gamma \rightarrow 1$, or a global optimization when $\gamma \rightarrow 0$.

In addition to evaluating n_i's neighbors with respect to their load and view, Quo Vadis introduces a shortest path component into the computation of n_j's overall utility. This way, Quo Vadis accommodates two otherwise contradicting metrics, namely load and distance. The shortest path component is of the form:

$$SPC_j = \frac{dist_{j,d}}{dist_{i,d}} * (v_i + \rho_i) \tag{6}$$

Assuming $dist_{i,d}$, v_i and ρ_i remain constant during the evaluation of n_j, SPC_j is smaller for neighbors that move the route closer to the destination. The overall utility of n_j can be determined by:

$$U_j = \beta * SPC_j + (1 - \beta) * LU_j \tag{7}$$

Here, the tuning parameter β determines the significance of the shortest path component in U_j. For two neighbor nodes $n_{j'}$ and $n_{j''}$ the shortest path component can compensate for a difference in load utility so as to guide the route through a neighbor with less favorable load promise yet better distance estimate.

$\beta \rightarrow 1$ will bias the utility computation towards a shortest Euclidean Distance, while $\beta \rightarrow 0$ would try to optimize the route with respect to the overall load. Clearly, there is a wide variety of possible routing decision functions which triggers the question of what constitutes a *good* one. As with knowledge representation, the choice of a particular decision strategy will always introduce bias. Its quality must be evaluated with respect to a given network topology.

3 Simulation and Preliminary Results

The experimental design of Quo Vadis involves the simulation of a very large communication environment [1]. Several experiments have been and are being conducted in a regular grid topology environment. Simulations of Quo Vadis with regular rectangular grid networks display a variety of promising results. These include an inherent load balancing behavior and the ability of nodes to establish routes so as to circumvent highly loaded network areas. The parameterized knowledge representation and heuristic routing functions used in Quo Vadis naturally cause congested or highly utilized areas to have a repulsive effect on new routes as they are being established. The extent of load balancing can be tuned by adjusting the choice of parameters in the heuristic routing functions.

We have compared the routing decisions made by Quo Vadis with decisions made by a fully informed optimization algorithm, namely the Bellman-Ford dynamic programming algorithm. Both strategies tried to establish routes between the same random source-destination pair so as to minimize the total node utilization.

Preliminary results indicate that Quo Vadis succeeded in finding *good* routes despite the fact that decisions are based upon summary information. Due to exhaustive search, the Bellman-Ford strategy always determined the *best* route between a source and a destination in a given network state. Routes established by Quo Vadis between the same source and destination encountered only marginally larger cumulative node utilization. However, Quo Vadis' placement of routes resulted in a better overall utilization of the network and lower average per-hop-delay.

In a simulation experiment we have established k random connections in two separate 10x10 node grid networks, using Quo Vadis and Bellman-Ford strategies respectively. Each connection was granted a random bandwidth requirement of b units, where $0 \leq b \leq b_{max}$. After k connections had been established, the per-hop-delays D_{QV} and D_{BF} were determined for Quo Vadis and Bellman-Ford strategy respectively. As newly established connections can significantly impact the delay of already existing routes, it is imperative for a good routing algorithm to not only base its decisions on a given network state but to implicitly anticipate complex network dynamics. The resulting per-hop-delay ratio $R = D_{QV}/D_{BF}$ depending on network load (simulated by increasing k) is shown in Figure 1. While routes established by the Bellman-Ford strategy display a smaller average per-hop-delay in a sparsely loaded network, Quo Vadis prevails as the network's load increases. The significance of these findings is underlined by the fact that Quo Vadis routing resulted in a shorter average route length.

4 Future Work

Quo Vadis offers an extremely challenging testbed for the state-of-the-art knowledge representation and machine learning algorithms. It uses parameterized knowledge representation and heuristic routing functions. The performance of Quo Vadis (measured in terms of metrics such as utilization, delay, length of route, etc.) is a function of parameters (α, β, γ). This

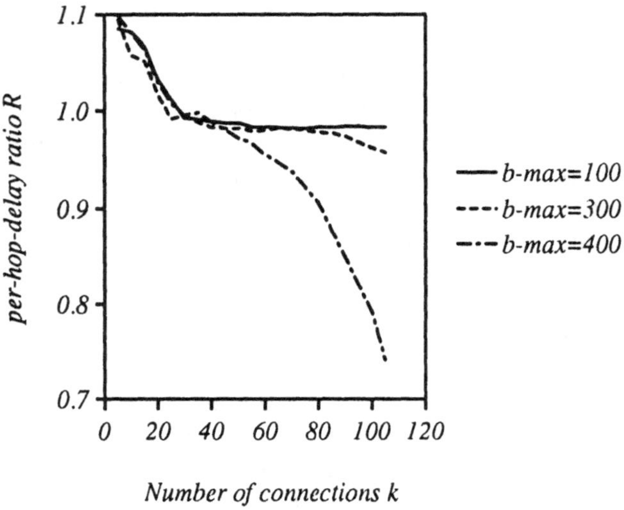

Figure 1: Effect of network load on avg. per-hop-delay ratio $R = D_{QV}/D_{BF}$

makes it possible to modify the behavior of Quo Vadis in response to different network dynamics by tuning the adjustable parameters. A variety of statistical and neural learning algorithms for doing this are currently under investigation.

References

[1] A.R. Mikler, V.G. Honavar and J.S.K. Wong, "Simulating a Traveller: A Heuristic Approach to Routing in Large Communication Networks", *Proc.Euro. Sim. Symp., ESS '92*, Nov.1992, pp. 297-301

[2] L. Kleinrock and F. Kamoun, "Hierarchical routing for large networks", *Computer Networks*, No.1, 1977, pp. 155-174

[3] P.F. Tsuchiya, "The Landmark Hierarchy: A new Hierarchy for routing in very large networks", *Proc. ACM SIGCOMM 1988*, pp. 35-42

[4] R. Perlman, "Hierarchical Networks and the Subnetwork Partition Problem", *Comp. Networks and ISDN Syst.* No.9, 1985, pp. 297-303

[5] A.R. Mikler, V.G. Honavar and J.S.K. Wong, "Quo Vadis – A Framework for Adaptive Heuristic Routing in Very Large Communication Networks", Tech. rep. in preparation, Dept. of Computer Science, Iowa State University.

Hopfield Optimization Techniques Applied to Routing in Computer Networks

Sabine Neuhauser

Aachen University of Technology, Department of Computer Science IV
D-52056 Aachen, Germany
sabine@informatik.rwth-aachen.de

Abstract

Hopfield networks are investigated as a means of implementing Quality of Service (QOS)-driven routing in multimedia computer networks. A new mapping of router networks onto continuous Hopfield networks is derived. Simulations for a network of 10 routers show the dependencies on parameters and the difficulties encountered.

1 Motivation

Multimedia routing means finding paths in a computer network for packets of different types of data by means of routing algorithms employing QOS-driven metrics (i.e. cost functions for the evaluation of paths). The QOS parameters characterized by the multimedia application which wants to transmit data determine the routing metric and consequently influence the choice of the route. Each application could be assigned its own application-specific metric yielding application-specific routes.

QOS-driven routing implies multiple routes for each destination depending on the requested type of service. As a consequence, the number of routes to be computed and stored rises. Therefore, ways to compute application-specific routes if needed instead of in advance have to be thought of. One possibility is to set up application-specific virtual connections for multimedia applications. To keep setup times low, a fast routing algorithm is needed. In this article, Hopfield networks are investigated as an adaptive and efficient implementation of QOS-driven routing.

2 Routing as a Hopfield optimization task

A Hopfield network consists of one layer of neurons interconnected by weighted arcs. A neuron i computes its output V_i^{new} depending on symmetric weights w_{ij} and its external input In_i using update rule (1).

$$V_i^{new} = g_\beta\left(\sum_{j,j\neq i} w_{ji} \cdot V_j + In_i\right),\ g_\beta(x) = \frac{1}{1+\exp^{-\beta x}}, \tag{1}$$

The neurons compute their new outputs according to (1) until a stable state is reached. [1] proves that the stable states minimize function (2).

$$E(V) = -0.5 \cdot \sum_i \sum_j w_{ij} \cdot V_i \cdot V_j - \sum_i In_i \cdot V_{ii} \tag{2}$$

2.1 Representation and mapping

A network of routers is represented by two neurons per existing link plus an additional neuron closing the loop from the destination to the source router (fig. 1)

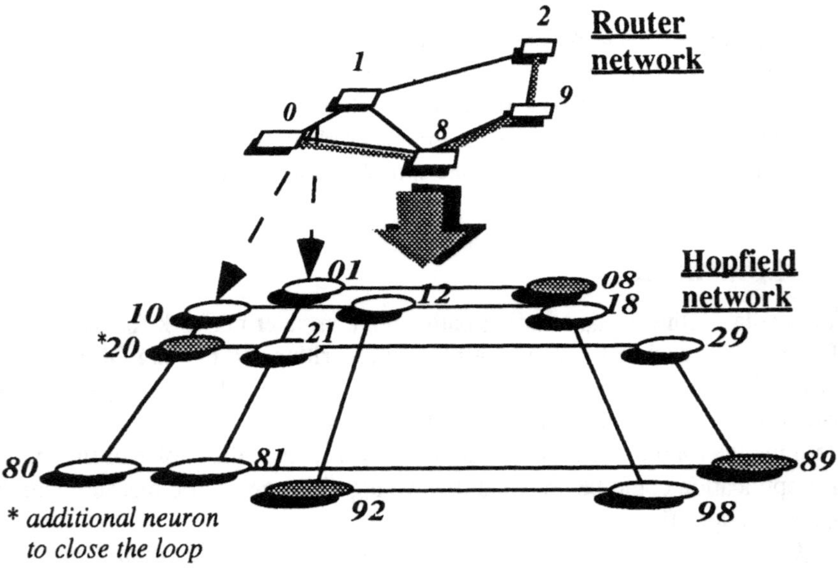

Fig. 1: Representing route 0-8-9-2 by a Hopfield network (inhibitory edges only)

In the stable state, the active neurons are mapped to a route according to (3), with x and y being routers and V_{xy} being the output of neuron n_{xy}. The simulations (chapter 3) have shown that $\delta=0.2$ is sufficient. If δ is chosen to be very small, more iterations are needed for equivalent results.

$$link(x,y) \in route \Leftrightarrow V_{xy} \geq 1-\delta, \qquad \delta > 0, \delta << 1 \tag{3}$$

The task of routing can be seen as the task of finding state variables V_{xy} such that the sum of distances along the path is minimized on condition that the path unambiguously connects the source router s with the destination router d by means of existing links. That is, the routing problem is formulated in terms of a constrained discrete minimization problem which can be transformed into an unconstrained optimization problem by means of penalty function terms. Defining (4), routing is mapped to the optimization problem of minimizing (5).

$$\overline{V_{xy}} = V_{xy} \cdot e_{xy} \quad with \quad e_{xy} = \begin{cases} 1 & if\_link(x,y)\_exists \\ 0 & else \end{cases} \tag{4}$$

$$R(V) = P1 \cdot \sum_{xy} d_{xy} \cdot \overline{V_{xy}} + P2 \cdot \sum_{x} \left(\sum_{y} \overline{V_{xy}} - \sum_{y} \overline{V_{yx}} \right)^2 + P3 \cdot \left(1 - \overline{V_{ds}}\right) \tag{5}$$

The penalty parameters $P1$, $P2$ and $P3$ stand for minimizing distances, guaranteeing unambiguous ways and closing the loop respectively. As in [3], the inputs (6) and weights (7) for the Hopfield network specifically designed to solve a routing problem for a given source-destination pair can be derived either plausibly, by comparison of coefficients [1] or by means of the equations of motion [4]. The representation is sensible as the weights turn out to be symmetrical. The distances d_{xy} contribute to the inputs only.

$$In_{xy} = \begin{cases} -P1 \cdot d_{xy} & if\_link(x,y)\_exists \\ P3 & if\_x == d\_and\_y == s \\ 0 & else \end{cases} \tag{6}$$

$$w_{xy,ij} = \begin{cases} 2 \cdot P2 & if\_x == j \\ 2 \cdot P2 & if\_y == i \\ -2 \cdot P2 & if\_x == i \\ -2 \cdot P2 & if\_y == j \end{cases} \tag{7}$$

2.2 Implementation

Simulating the derived Hopfield network means that a source-destination pair is given as an input into the Hopfield network, the outputs are initiated and new outputs are computed until the network settles to a stable state representing the requested route. The main problems in programming an appropriate software simulation of Hopfield networks are the choice of the penalty parameters and of the update rule.

The software implementation aimed at simulating a continuous update. However, a router implementing a Hopfield network by permanently solving differential equations would not be feasible. An easier approach based on the first Euler method has been taken [4]. In each iteration, all neurons simultaneously are updated according to (8).

$$V_{xy}^{new} = V_{xy} + \varepsilon \cdot \left(g_\beta \left(\sum_{ij} w_{ij,xy} \cdot V_{ij} + In_{xy} \right) - V_{xy} \right), \quad \varepsilon \in]0;1] \tag{8}$$

with ε being a smoothing constant. If $\varepsilon<1$, the change of the output of one neuron almost immediately contributes to the changes resulting in the activities of neighboured neurons. Thus, the system is forced to slowly converge towards the stable state rather than to oscillate and the differential equations governing the system are approximated.

2.3 Comparison to other approaches of Hopfield routing

A similar approach of [2] employs two additional penalty parameters, for deactivating neurons corresponding to non-existent links and for guaranteeing convergency towards activities of 0 or 1. Moreover, the update formula (8) yields better convergency results in comparison with [2].

[4] proposes to use $\varepsilon = 1$ which did not work in this approach. His hill-climbing terms would induce a more complicated representation of the problem.

3 Experimental results with Hopfield Routing

Experiments have been carried out with a network of 10 routers and 17 links represented by a Hopfield neural network of 35 neurons which has been simulated by software based on the update rule (8). Each experiment consisted of finding all 100 routes in this network of routers. The simulation parameters which have to be fixed are the penalty parameters P1, P2 and P3 as well as β in (1) and ε in (8). 19 sets of parameters P1, P2 and P3 (called PSi) have been tested. The results shown were obtained by choosing $\varepsilon = 0.2$ and $\beta = 1.0$.

3.1 Results

The performance of the Hopfield network strongly depends on the set of penalty parameters chosen. There are some sets such as PS2 (P1=275, P2=1000, P3=1225) or PS12 (P1=2.5, P2=10, P3=12.5) which turned out to be very successful in finding nearly 100 % of the routes (fig. 2) as well as in converging very quickly (fig. 3). Yet, low iteration numbers do not imply that routes are really found. For example, with PS6 (P1=12.5, P2=15, P3=20) the network converges almost immediately, but mostly the final state is not admittable.

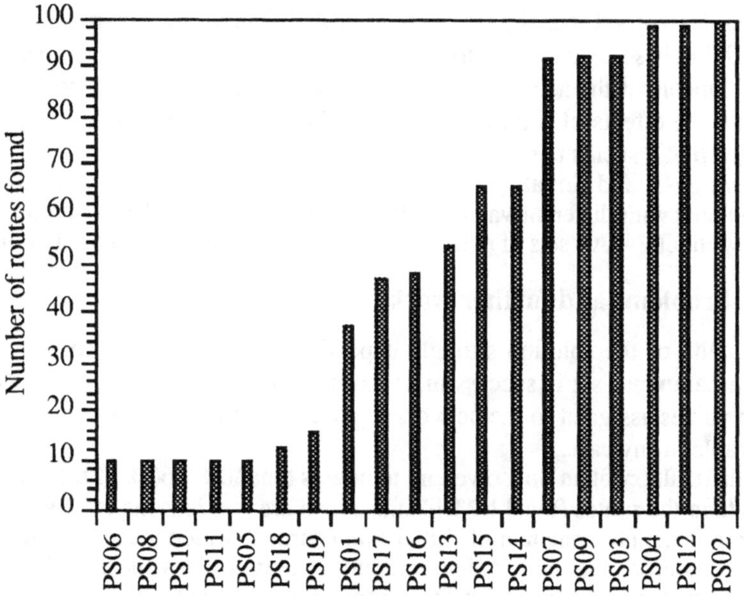

Figure 2: Number of routes found for different sets of penalty parameters PSi

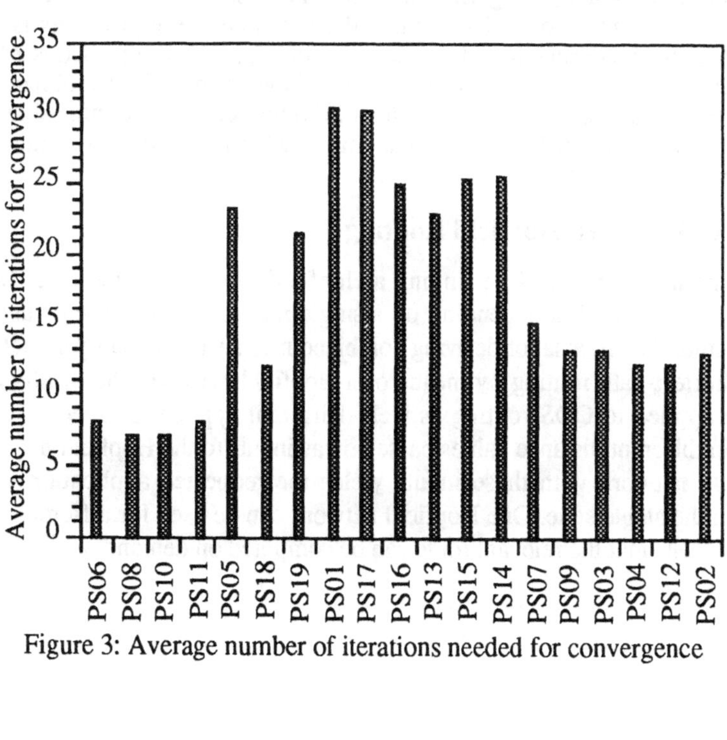

Figure 3: Average number of iterations needed for convergence

Additional experiments with networks of 50 and of 100 routers showed that the number of iterations needed seems to be independent of the problem size, as [4] already indicated. An ε of 0.2 is the most obvious one to choose for the update formula. Even good sets of parameters such as PS2 performed extremely bad for an $\varepsilon>0.3$. These results support the use of the update formula (8). That is, a smooth update of the activities is a prerequisite for the network to converge. In contrast, the effects of β on the quality of the solution are difficult to predict, though a β of 1 seems to be sensible.

Kamoun's additional penalty parameter P4 [2] was not needed, as experiments with different values of P4 have been showing. PS2 even performs better without P4. All sets of parameters and more figures are to be found in [3].

3.2 Problems and further work

The quality of the solution strongly depends on the parameters to be chosen. Once a parameter set is successful, it has shown to be successful for different metric values assigned to the network of routers though a 100%-solution is not achievable in any case.

A difficult problem not solved up to now is that 40% - 60% of the routes are worse than the routes found by Dijkstra. There are two reasons for that. First of all, there is the phenomenon of "killer link neurons" which means that there are activated neurons that build a detour around a neuron. These neurons hinder the neuron that takes the short cut from becoming activated. Secondly, there are superfluous neurons giving rise to additional closed routes debilitating the neurons on the main route. One of the ideas to solve these problems is to add guard neurons hindering those killer link neurons from becoming activated.

Finally, though the number of iterations is independent of the problem size, a sequential simulation of more than about 20 links seems to be infeasible as the number of neurons needed grows then. That is, a hardware solution would be the only feasible solution.

3.3 QOS-driven Hopfield Routing

The experiments focused on finding a Hopfield implementation of a shortest-path routing algorithm depending on a single metric, the distance of the links. Connecting the proposal of deriving combined metrics from QOS parameters [3] with shortest-path routing by means of a Hopfield network, the results gained can be applied to QOS-routing as well. Different application-specific metrics result in different distance values passed on as inputs to the Hopfield network. A Hopfield network with these inputs yields the requested application-specific route in the stable state. One Hopfield network can be used for different metrics which means that the relevant route can be computed on demand.

4 Conclusion

Hopfield neural networks have been presented as an implementation of a routing algorithm that could be called as and when required thus relieving routing tables in the face of multiple routes needed for QOS-driven routing. An implementation for a Hopfield neural network shows considerable results. Neural routing in large networks only is feasible, if neural hardware is available. The work is part of a project on QOS-driven routing [3].

Acknowledgements

This work is supported by the Graduiertenkolleg "Informatik und Technik" (Graduate College "Methods and tools of computer science and their application in technical systems") at Aachen University of Technology.

References

[1] J.J. Hopfield, D.W. Tank: "Neural Computation of Decisions in Optimization Problems", Biological Cybernetics 52, 141-152, 1985

[2] F. Kamoun, M. Ali: "A Neural Network Shortest Path Algorithm for Optimum Routing in Packet-Switched Communications Networks", Globecomm '91, Phoenix, Arizona, 12/91

[3] S. Neuhauser: "QOS-driven routing in multimedia communication networks" in *Projects of the Graduiertenkolleg "Informatik und Technik"*, Aachener Informatik-Berichte 93-7, RWHT Aachen, 1993

[4] Y. Takefuji: *Neural Network Parallel Computing*, Kluwer Academic Publishers, 1992

ATM Call Control by Neural Networks

Joaquim E. Neves*, Luis B. de Almeida and Mário J. Leitão*****

** DEI, Universidade do Minho, Azurém, P-4800 Guimarẽs*
*** DEEC-IST, Avenida Rovisco Pais, P-1000 Lisboa*
**** DEEC-FEUP, Universidade do Porto, Rua dos Bragas, P-4000 Porto*
INESC - Instituto de Engenharia de Sistemas e Commputadores
,* Largo Mompilher 22, P-4000 Porto*
*** Rua Alves Redol 9, P-1000 Lisboa*
eneves@inescn.pt, lba@sara.inesc.pt, mleitao@inescn.pt

Abstract

The resource allocation in the Broadband Integrated Services Digital Network (B-ISDN) must guarantee the quality of service negotiated with new and existing calls, taking into account the Asynchronous Transfer Mode (ATM) statistical characteristics. A quality of operation function, characterizing the overall network performance, is proposed, and based on this function, it is introduced a new strategy for the admission control and routing of the ATM call connections. As it is shown by simulation results, feedforword Neural Networks trained with the backpropagation algorithm, can learn the traffic patterns in previous traffic situations, and can be used to predict the quality of operation changes caused by each new call.

1 Introduction

The transport of B-ISDN services, like interactive video telephony, high quality video and audio broadcast programs or data file transfer, requires a call establishment phase. The Asynchronous Transfer Mode provides the transport and the switching of information flow, generated by B-ISDN services, in fixed size data packets called ATM cells which can be transported in the existent Plesiochronous (PDH) or Synchronous (SDH) Digital Hierarchies or in the new cell-based transmission systems [1].

At the call establishment, the user has to negotiate with B-ISDN control entity the traffic characteristics of the call and the quality of service requested. The network control entity can accept the request and allocate network resources for the support of the service, or propose a lower quality of service, and in the limit, rejected the call, if not enough network resources are available. B-ISDN has mechanisms for usage parameter control by policing the call traffic at user interface and taking appropriate actions if the usage values of the information flow parameters, namely, the average and the peak cell rate and the burstiness, are exceeded in a virtual channel.

Neural networks have been proposed to control ATM networks. For example in reference [7], three hierarchical levels of neural networks are suggested to implement the control functions. Neural networks can also be applied in service coding [9] (e.g. video and audio compression), in the B-ISDN policing function, (e.g. the control of a selective cell discard), in switching control [2, 8] and in the processing of the operation and maintenance (OAM) signals. Neural Networks can learn traffic patterns of the network operation in previous situations, to be used in the cost prediction of each new call admission.

This paper discusses the application of neural networks in ATM call control. Section 2 introduces the network quality of operation function and describes its use for call control proposes. Section 3 describes and compares simulation results with different call admission control methods. Section 4 summarizes the main issues of this paper.

2 ATM Call Control Technique

The resource allocation for ATM call connections can be made by taking the peak bit rate of ATM cells sources as reference, but no statistical gain is obtained by multiplexing many sources. If the resource allocation is made by the average bit rate of ATM cell sources, the statistical gain obtained by multiplexing many sources is maximum but the simultaneous occurrences of the peak periods of some sources can drastically increase the delay and cell loss rate. The strategy proposed for call admission control establishes a compromise between the maximum number of calls accepted and the satisfaction of the quality of services negotiated for calls established.

2.1 Quality of Operation of the B-ISDN

The Quality of Operation concept integrates the parameters of the quality of service negotiated by the network in the call establishment, the availability of network resources, and the equilibrium between the call rejection rate of different ATM service classes. The proposed quality of operation function (QO) can be written by the following expression:

$$QO = \sum_j (\alpha_j A_j + \beta_j B_j - \chi_j X_j - \sum_i \delta_{ji} \Delta_{ji}) \tag{1}$$

where α_j, β_j, χ_j, and δ_{ji} are non-negative real control parameters, and A_j, B_j, X_j, and Δ_j are functions. A_j quantifies in terms of quality of operation the bit rate allocated to each service class j; B_j quantifies the bit rate free to be allocated to each service class j; and X_j quantifies the deviation of the call rejection rate of the service class j from the average call rejection rate of all service classes; Δ_{ji} quantifies the main quality of service requirements of each service class j, namely the cell loss rate (i=0), the delay (i=1) and the delay variation (i=2).

The values of the control parameters are dependent of the B-ISDN operation scenarios, the predominant services and the desirable network load. The variables of the quality of operation functions can be average or accumulated traffic values. The acquisition times of these variables have to be compatible with the time constants of the services and network.

The bit rate allocated function A_j is taken as a nonlinear function of the allocated bit rate to the service class j, with an increasing contribution to quality of operation if the allocated bit rate to that service class does not reach a certain threshold, and with a decreasing contribution if the allocated bit rate is bigger than that value. This allows the dynamic partition of band between service classes.

If a static partition is desired, the sum of the threshold values of the bit rate allocated to each service class j is smaller than the transmission capacity; in this case the bit rate free to be allocated function B_j and the deviation of the calls rejection rate function X_j do not need to be considered ($\beta_j = \chi_j = 0, \forall j$).

The B_j and X_j functions produce opposite effects, which allows that in heavy traffic situations, the resources allocated to narrowband services do not block the access to broadband services. In reference [6], the proposed solution for this problem is static partition of band.

The quality of service functions express the contribution of the corresponding factors (the cell loss rate Δ_{j0}, the delay Δ_{j1}, and the delay variation Δ_{j2}) to the quality of operation. The values of the Δ_{ji} functions for each service class j, are proportional to the fraction of the bandwidth used by the class.

2.2 Traffic Prediction by Neural Networks

Patterns of the traffic load in a node or link can be collected during the operation of the B-ISDN, in different traffic situations, to be used as learning patterns of neural networks. The delay and the cell loss rate that will be introduced by the call are not known at the time of the call establishment. When the resources of the new call connections are established, the vector of the allocated bit rate of each service class has to be stored and the traffic load pattern is evaluated later, when the new call is generating traffic.

Neural Network inputs are the allocated bandwidth to each service class, and the outputs can be the expected delay, cell loss rate, and the maximum and minimum buffer occupation, the latter leading directly to the delay variation. Another output is included (the number of arrived cells) to allow a better behavior of the training process.

2.3 Call Admission Control

When a connection request for resource allocation to a call arrives to a node, the bit rate requested and the maximum delay and error rate are declared, and the connection is integrated in a suitable service class. The decision if the connection request can be accepted or has to be rejected is based in the network quality of operation expected in each node of the call route.

Each node processor asks the neural network the expected traffic load pattern for the node and the outgoing link, with and without the inclusion of the new connection. The network answers with the expected patterns, then the quality of operation is evaluated in both cases, and the resources are allocated to the call if the expected quality of operation in every B-ISDN node and link of the call route is higher if the new connection was accepted.

2.4 Routing of the ATM Calls

In case where a call has available alternatives routes, every network node and link of all available call paths is inquired about the network quality of operation expected for the call. This allows to discard unacceptable routes.

The cost function of the routing algorithm can be obtained from the quality of operation expected in the nodes and in the links of each call path, with suitable values of the control parameters. For instance, to route calls through paths less loaded, the control parameters of the allocated bit rate have to be set with a small or even null value.

If the number of alternative routes, and the number of nodes of each route, is small, the best path of each call can be found in real time for each call, otherwise the best routing can be determined periodically and all calls within the same time interval follow the established route.

3 Performance Simulation

The B-ISDN components (transmission links and switching nodes), traffic sources and procedures (routing and flow control) are simulated according to the model presented in reference [4], while the ATM traffic is described and simulated by the Markovian model described in reference [3]. Transmission links and switching nodes are represented by delay, error rate, throughput and buffer length. Each ATM traffic source is defined by two Markovian state space processes. The call birth of different services and users are calculated by one Markovian process with different duration and average time between call birth, in each state. The cells of each call are generated, with the appropriate parameters for each service, by the other state space process of the same traffic model. Table 1 shows the peak and the average cell rate and the burstiness of the service classes used in our simulations. The minimum interval between cells is taken equal to the simulation time unit (2.7 μs, for 53 byte cells at 155.52 Mbit/s).

Table 1: Main service characteristics

Service Class (SC)	**Cell Rate (KCell/s)**		**Burst (%)**
	Average	**Peak**	
SC_0	1.604	10.000	13.36
SC_1	3.750	5.000	50.00
SC_2	20.000	20.000	100.00

The prediction of the traffic patterns for each call has been made by a 3 layer neural network with 7 neurons in the hidden layer, and a hyperbolic tangent activation function in the internal neurons. The neural network has been trained with 3500 traffic patterns generated by timing reports produced, with an interval of 5 ms, during previous simulations of the B-ISDN operation. For training the neural network, the backpropagation algorithm was used with adaptive learning rate parameters [5] and the sum of squared errors as cost function.

Figure 1 shows the allocated bandwidth during the 20 seconds of simulation time in one node of a network loaded by calls from the services introduced in table 1. The results are normalized to the node capacity and are shown in four cases: (a) without rejection; (b) allocation based on average cell rate; (c) allocation based on the peak cell rate; (d) allocation based on the proposed technique, with the following values for the control parameters: $\alpha_j = 1.0$, $\delta_{j1} = 0.3$, and $\beta_j = \chi_j = \delta_{j0} = \delta_{j2} = 0.0$, $\forall j$. As seen in the figure if the resource allocation is made without call rejection, an overflow is reached in most of the simulation time. With the allocation by the average and the peak cell rate only the narrowest band service class (narrowest average and narrowest peak, respectively) can access the network resources, namely during the significantly loaded periods. With the proposed technique, the figure shows that all the service classes can share the available resources even when demand is higher.

Table 2 shows the effectiveness of the model to control the delay in the

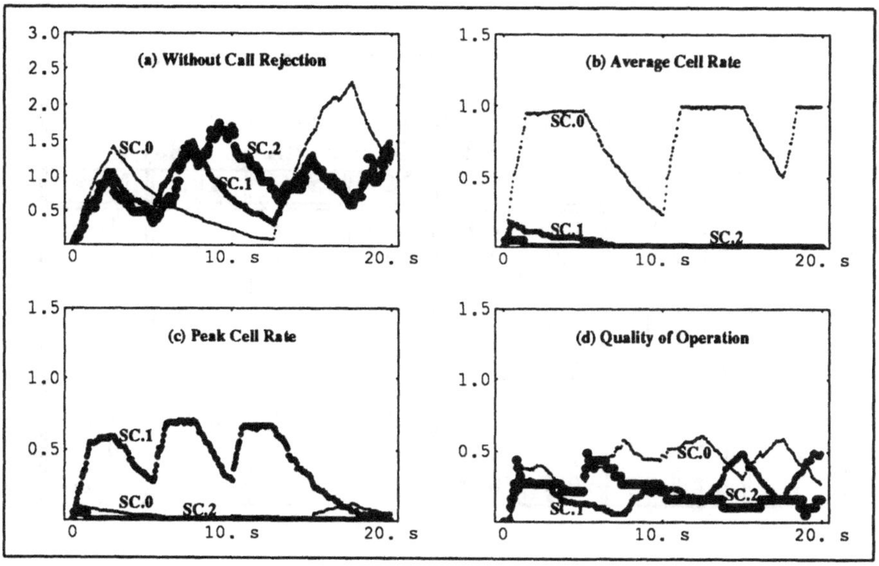

Figure 1: Normalized allocated bandwidth to each service class

network, by adjusting properly the delay control parameter. The output of the simulation is presented in tabular form for nine cases; for each one, the results shown are the normalized maximum allocated bit rate to each service class, and the average and maximum delay, normalized to the maximum buffer occupancy, obtained during the 20 seconds of simulation time. The network was loaded by calls from the services characterized in table 1, and generated with the same statistical distributions of the simulations reported in figure 1. Cases (a), (b), (c) and (d) of table 2 correspond to the simulation presented in figure 1. In cases (a) and (b), the table shows that a full buffer situation is reached (and indeed overflow!); in case (c) no overflow occurs, but the network is lightly loaded due to the peak cell rate allocation; case (d) gives good control on the delay. In case (e) and (f) the control parameter is increased with respect to case (d), which causes a corresponding reduction in the delay. Finally in the last three cases, the delay control parameter is significantly reduced for each of the three service classes. The results show that the network maximum delay found in the node is controlled in case (h) and (i), but not in (g). The explanation for this is the fact that in (g), service class 0 was lightly controlled with respect to the delay, and being a bursty service, it escapes from the network control capabilities. It is noted that other control parameters of the quality of operation function could be effective to control the delay, but in an indirect way, such as cell loss and delay variation.

Table 2: Call control: Allocated bandwidth and delay

Case	**Delay Control Parameter**			**Normalized Max. Alloc. Bandwidth**			**Normalized Delay (%)**	
	δ_{01}	δ_{11}	δ_{21}	**SC_0**	**SC_1**	**SC_2**	**Ave.**	**Max.**
a	0.0	0.0	0.0	2.32	1.48	1.74	0.96	1.00
b	0.0	0.0	0.0	0.99	0.17	0.05	0.53	1.00
c	0.0	0.0	0.0	0.13	0.71	0.05	0.07	0.19
d	0.3	0.3	0.3	0.61	0.50	0.49	0.36	0.85
e	0.5	0.5	0.5	0.52	0.49	0.38	0.26	0.74
f	0.7	0.7	0.7	0.63	0.67	0.33	0.27	0.64
g	0.1	0.7	0.7	0.89	0.31	0.38	0.46	1.00
h	0.7	0.1	0.7	0.79	0.64	0.33	0.28	0.59
i	0.7	0.7	0.1	0.55	0.64	0.16	0.33	0.68

4 Conclusions

The Quality of Operation incorporates the total amount of traffic, the cell loss rate, the delay and the delay variation, as well as the availability of network resources related to the free transmission capacity that can be allocated to each service, and the equilibrium between the call rejection rate of different ATM services. During the operation of the B-ISDN in different traffic situations, patterns of the traffic load in the node or link are memorized to be used as training patterns of a neural network.

When a request for resource allocation to a call arrives to the B-ISDN control entity, the neural networks associated to the B-ISDN nodes and links of the path are inquired about the traffic patterns expected with and without the new connection, and the resources are allocated to the call if the overall quality of operation is expected to increase if the new connection is accepted. Simulation results show that the use of neural networks within this technique lead to a high flexiblity, allowing the control of network parameters according to specific service needs.

References

[1] ., "B-ISDN draft recommendations," in *Rec. I Series*, Geneva: CCITT, May 1990.

[2] T. X. Brown, "Neural networks for switching," *IEEE Communications Magazine*, vol. 27, no. 11, pp. 72–80, Nov. 1989.

[3] J. E. Neves and M. J. Leitão, "A markovian model for ATM traffic generation," to appear in Malaysia International Conference on Communications - MICC 93, Nov. 1993.

[4] J. E. Neves and M. J. Leitão, "Modeling of B-ISDN for performance simulation of ATM services," in *International Symposium on Modular Information Systems and Networks - ICSNET 93*, (St. Petersburg), Sept. 1993.

[5] F. M. Silva and L. B. Almeida, "Acceleration techniques for the backpropagation algorithm," in *Neural Network EURASIP Workshop Proceedings*, (Springer-Verlag), 1990.

[6] E. D. Sykas, K. M. Vlakos, I. Venieris, and E. N. Protonotarios, "Simulative analysis of optimal resource allocation and routing in IBCN's," *IEEE Journal on Selected Areas in Communications*, vol. 9, no. 3, Apr. 1991.

[7] T. Takahashi and A. Hiramatsu, "Integrated ATM traffic control by distributed neural networks," in *XIII Internation Switching Symposium*, vol. III, (Stockholm), May 90.

[8] T. Troudet and S. Walters, "Neural network architecture for crossbar switch control," *IEEE Transactions on Circuits and Systems*, vol. 38, pp. 42–56, 1991.

[9] Yi-Tong, R. Chellapa, A. Vaid, and B. K. Jenkins, "Image restoration using a neural network," *IEEE Trans. on Acoustics Speech, and Signal Processing*, vol. 36, no. 7, pp. 1145–1151, July 1988.

Channel Equalization by Distribution Learning: The Least Relative Entropy Algorithm

M. Kemal Sönmez and Tülay Adalı*

Department of Mathematics, University of Maryland
College Park, MD 20742
kemal@src.umd.edu
**Department of Electrical Engineering, University of Maryland*
Baltimore, MD 21228-5398
adali@engr.umbc.edu

Abstract

We formulate the adaptive channel equalization as a conditional probability distribution estimation problem. Conditional probability density function of the transmitted signal given the received signal is parametrized by a sigmoidal perceptron. In this setting, the natural choice for a cost function is the relative entropy (Kullback-Leibler discriminant) between the desired probability density function and the perceptron model. Since the desired probabilities are not readily available for a given channel, we use first order stochastic approximations. The validity of such an approach in terms of consistency and asymptotic behavior is shown by linking the scheme to maximum *partial* likelihood estimation. We use gradient descent on this relative entropy cost function to derive the Least Relative Entropy (LRE) algorithm. The complex LRE algorithm is applied to the equalization of multipath channels, and is shown to outperform the widely used Least Mean Square (LMS) algorithm with a computational complexity comparable to that of LMS.

1 Introduction

Adaptive equalization techniques developed during the last two decades have been successfully implemented in high speed data transmission for correcting the distorting effect of channels having substantial amplitude and delay distortion. Traditionally, channel equalization is considered as a deconvolution problem where an equalizer is constructed such that the impulse response of the channel and equalizer combination is as close to a delta function as possible (with a constant delay d). In this work, a new formulation is presented for the adaptive channel equalization problem in which a parametrized model of the conditional distribution function of the transmitted signal given the received signal is estimated. A sigmoidal perceptron is used as the probability distribution model, thus parameter estimation is accomplished by learning from training data on the perceptron.

The organization of the paper is as follows: Section II presents the problem formulation and the model for probability distribution. In Section III, we introduce the entropic cost function we use with this formulation, and present the derivation of the Least Relative Entropy (LRE) algorithm for binary signals. The derivation is then extended to complex bipolar signaling. In Section IV, the complex LRE algorithm is applied to the equalization of multipath channels, and is compared with the normalized LMS algorithm.

2 Probability Distribution Model

Consider the adaptive channel equalization problem shown in Figure 1 in which the probability that the transmitted signal $x(n)$ is equal to a_i from the S-symbol source alphabet given the finite past of the received signal $\mathbf{y}(n) = [y(n), y(n-1), ..., y(n-N+1)]^T$ is to be determined from a training sequence. The sigmoidal perceptron $p_{\mathbf{w}}$ parametrizes the conditional probability distribution in the following manner: A sigmoidal perceptron is a function

$$p_{\mathbf{w}} : \mathbf{R}^N \longrightarrow [0,1]^K$$

which is defined as

$$\begin{aligned} p_{\mathbf{w}}(\mathbf{y}(n)) &\equiv p_{\mathbf{w}}(n) \\ &= \left[\frac{1}{1+e^{-[\mathbf{w}^T\mathbf{y}(n)]_1}}, \frac{1}{1+e^{-[\mathbf{w}^T\mathbf{y}(n)]_2}}, \dots, \frac{1}{1+e^{-[\mathbf{w}^T\mathbf{y}(n)]_K}}\right]^T. \end{aligned}$$

The $N \times K$ weight matrix $\mathbf{w}$ contains the parameters of the distribution and $K = \lceil \log_2 S \rceil$ is the number of perceptron outputs. The goal is to learn the probability distribution

$$P(x(n) = a_i | \mathbf{y}(n)) \approx p_{\mathbf{w}}(n)$$

by adjusting the parameters $\mathbf{w}$.

3 Least Relative Entropy Algorithm

3.1 Relative Entropy Cost Function

In the probability distribution learning framework presented in the previous section, the natural choice for a cost function is the relative entropy between the desired distribution function and its perceptron model [1]. The desired probabilities are in general not available, therefore we use first order stochastic approximations [2]:

$$p^{true}(n) = E[x(n)] \approx x(n)$$

which results in the Approximate Relative Entropy (ARE) given by:

$$\text{ARE}(\mathbf{w}) = -\sum_{n=1}^{M} x(n) \ln p_{\mathbf{w}}(n) + (1 - x(n)) \ln(1 - p_{\mathbf{w}}(n)). \qquad (1)$$

With this approximation, the ARE cost function becomes equivalent to the negative of *partial* log-likelihood [2], a factorization of full likelihood, which is a relatively recent method in estimation theory to extract information about parameters of interest in dealing with dependent data and to avoid nuisance parameter complications, [3, 4]. Partial likelihood is particularly suitable for time series applications, and in [5], it is shown that the maximum partial likelihood estimator using logistic regression model $p_{\mathbf{W}}$ is consistent and asymptotically normal. Furthermore, the ARE is a convex function, therefore if a perceptron model exists, it is unique.

In [6], it is shown that the gradient descent is guaranteed to find a solution provided that there is one, if the cost function is *well-formed.* A *well-formed* cost function is defined as a function whose gradient magnitude is always larger than a constant. The quadratic cost function is not well-formed in this sense and gradient descent learning on this cost function can not always find a viable solution. The entropic cost function given in (1) is *well-formed* in this sense, and is therefore guaranteed to find a solution.

3.2 The LRE Algorithm

Gradient descent minimization of the ARE cost function (1) results in the Least Relative Entropy (LRE) algorithm:

- Estimate the probability that $x(n) = 1$:

$$p_{\mathbf{W}}(n) = \frac{1}{1 + e^{-\beta \mathbf{W}(n)^T \mathbf{y}(n)}} \qquad (2)$$

$\min_{\mathbf{w}_I,\mathbf{w}_Q} \text{ARE}(\mathbf{w}_I,\mathbf{w}_Q) = -\sum_{n=1}^{M} \frac{1}{2}[(1+x_I(n))\ln(1+p_{\mathbf{w}_I})+$ $\frac{1}{2}(1-x_I(n))\ln(1-p_{\mathbf{w}_I}) + \frac{1}{2}(1+x_Q(n))\ln(1+p_{\mathbf{w}_Q})+$ $\frac{1}{2}(1-x_Q(n))\ln(1-p_{\mathbf{w}_Q})]$
Compute the probabilities that $x_I(n)=1$ and $x_Q(n)=1$ $p_{\mathbf{w}_I}(n) = \tanh[\beta(\mathbf{y}_I(n)^T\mathbf{w}_I(n) - \mathbf{y}_Q(n)^T\mathbf{w}_Q(n))]$ $p_{\mathbf{w}_Q}(n) = \tanh[\beta(\mathbf{y}_I(n)^T\mathbf{w}_Q(n) - \mathbf{y}_Q(n)^T\mathbf{w}_I(n))]$ Update the filter coefficients: $\mathbf{w}_I(n+1) = \mathbf{w}_I(n) + \alpha[(x_I(n) - p_{\mathbf{w}_I}(n))\mathbf{y}_I(n) + (x_Q(n) - p_{\mathbf{w}_Q}(n))\mathbf{y}_Q(n)]$ $\mathbf{w}_Q(n+1) = \mathbf{w}_Q(n) - \alpha[(x_I(n) - p_{\mathbf{w}_I}(n))\mathbf{y}_Q(n) + (x_Q(n) - p_{\mathbf{w}_Q}(n))\mathbf{y}_I(n)]$

Table 1: The Complex LRE Algorithm

- Update the pdf parameters:

$$\mathbf{w}(n+1) = \mathbf{w}(n) + \alpha(p_{\mathbf{w}}(\mathbf{y}(n)) - x(n))\mathbf{y}(n) \tag{3}$$

The complex version of the LRE algorithm and its cost function are given in Table 1. Note the use of a symmetric activation function for this case. The probabilities can be obtained by the transformation $\frac{1}{2}[(\cdot)+1]$.

4 Application to Multipath Channel Equalization

The complex LRE algorithm given in Table 1 is applied to the equalization of multipath channels in the presence of additive white Gaussian noise. A QPSK data transmission system is simulated in the block simulator Capsim environment [7]. The topology used in the simulations is shown in Figure 2. The additive white Gaussian multipath channel introduces severe distortion resulting in a *closed eye* situation. A sample received scattering diagram is shown in Figure 3. The LRE algorithm produces a probability map upon which a maximum likelihood decision is made. The probability map, given in Figure 4, is quite decisive for a large majority of received samples, hence the clusters at the highly probable regions of the map for the 4-element source alphabet.

The performance of the LRE algorithm is compared to that of the power normalized LMS algorithm. It is observed that the LRE algorithm which has complexity comparable to that of the widely used LMS algorithm has outperformed LMS at various channel distortion levels, especially at high multipath distortions (low main signal to multipath signal power ratios). Figure 5 shows a comparison of the bit error rates (BERs) for the complex LRE and LMS algorithms at various signal to noise ratios and at a main signal to multipath signal power ratio of 1.5. Each point on the curve is the ensemble average of 40 independent runs and the LRE shows an improvement of 1-6 dB in BER over what has been achieved by the LMS.

References

[1] J. J. Hopfield, "Learning algorithms and probability distributions in feed-forward and feed-back networks," *Proc. Natl. Acad. Sci. USA*.

[2] M.K. Sönmez, and J. S. Baras, "Time series modeling by perceptrons: A likelihood approach," *Proc. World Congress on Neural Networks*, (Portland, Oregon), pp. 601-604, July 1993.

[3] D. R. Cox, "Partial Likelihood," *Biometrika* vol. 62, no. 2, pp. 269-276(1975).

[4] W.H. Wong, "Theory of Partial Likelihood," *The Annals of Statistics*, vol. 14, no. 1, pp. 88-123, 1986.

[5] E. Slud, and B. Kedem, "Partial likelihood analysis of logistic regression and autoregression," Report TR91-47/ES/BK, Department of Mathematics, University of Maryland, College Park, 1991.

[6] B.S. Wittner and J. S. Denker, "Strategies for teaching layered networks classification tasks," *Neural Information Processing Systems*, (Denver, CO), pp. 850-859, 1988.

[7] T. Adalı, S. H. Ardalan, A.S. Sadri, "Multichannel general order FTF algorithm and its application to the equalization of mobile communication channels," in *Proc. IEEE Int. Conf. Acoust., Speech, Signal Processing* (San Francisco, CA), pp. IV.517-520, March 1992.

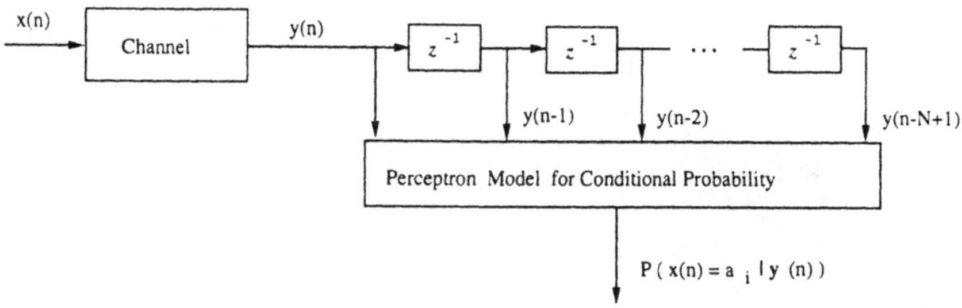

Figure 1: The Adaptive Equalization Problem

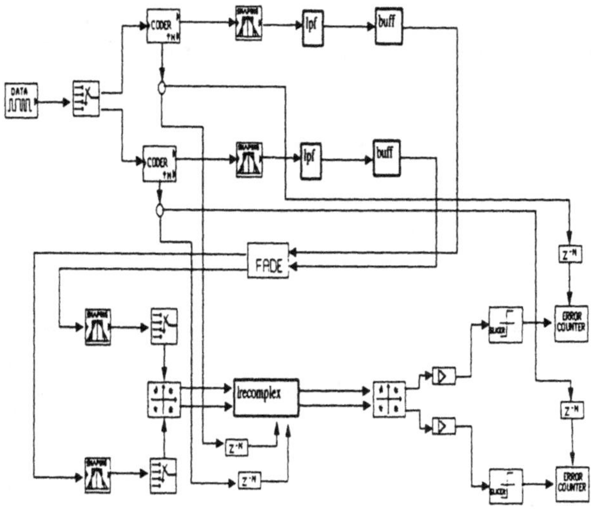

Figure 2: Capsim Topology for QPSK Data Transmission System

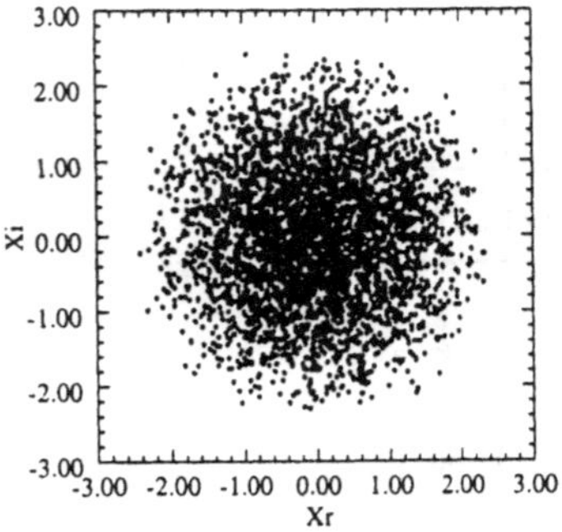

Figure 3:
Received QPSK Signal Constellation Diagram

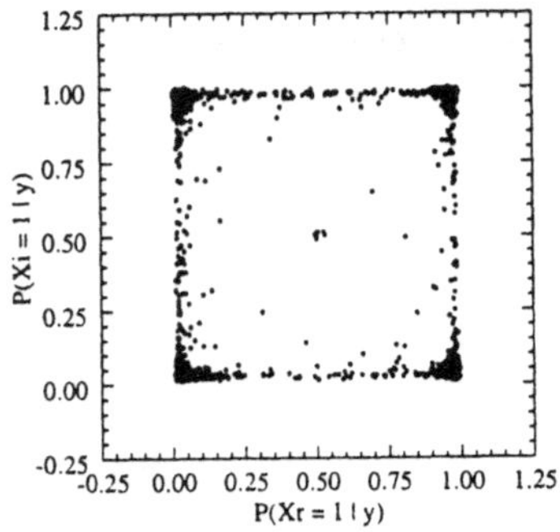

Figure 4:
Estimated QPSK Signal Probability Map

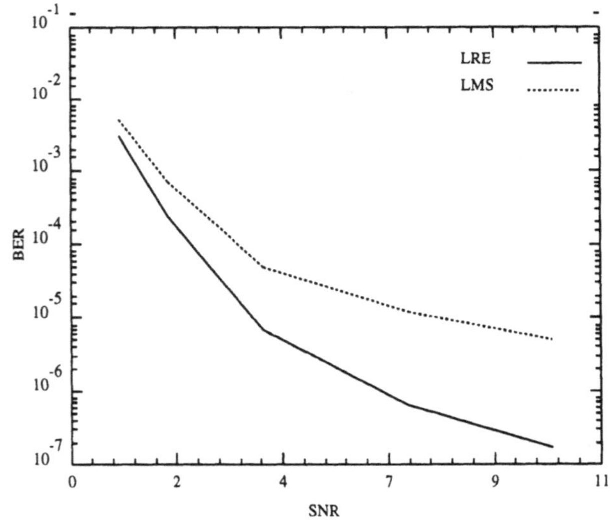

Figure 5:Bit Error Rate Comparisons for the LMS and LRE Algorithms
(Main signal power to multipath signal power = 1.5)

Optimal Traffic Routing Using Self-Organization Principle

Chen-Xiong Zhang

MAZ GmbH (Mikroelektronik Anwendungszentrum), Karnapp 20, D-21079 Hamburg, Germany
cxz@maz-hh.de

Abstract

This paper presents a new method to determine optimal traffic routing for telecommunication networks in a self-organizing way by means of an adaptive neural mapping. A traffic routing with minimal loss such as expected delay and other functions of traffic, is attained. The method can be effectively applied in large-scale communication networks.

1 Introduction

Traffic routing is an important task for telecommunication systems. Usually, communication companies choose traffic routes with minimal delay by using routing tables and dynamic routing [1][2][3]. However, in large telecommunication networks, the amount of traffic and the number of nodes and links are so large that the conventional routing algorithms are not effective enough to be implemented.

It is known that the Kohonen's self-organizing neural network [4] is based on a neural somatotopical mapping and used for the traveling salesman problem with success [5]. In the paper a new method based on modified Kohonen's model is presented to solve the large optimal routing problem in telecommunications in a more effective way.

2 Mapping Model for Traffic Routing

The routing problem has some similar characters to the multiple travelling salesmen problem which can be described as follows: each salesman travels along a subtour of customers. All subtours share a common depot, considered as a customer. Every customer, except for the depot, must be visited once by only one salesman so that the total distance travelled by all salesmen is minimum [6]. Here the subtour can be related to the connection paths and customer to the traffic node in the routing problem. However, in the routing problem the traffic node does not need to be used once through only one information transmission and the traffic capacity of each node is limited.

With the somatotopical mapping, a neural routing model is suggested in which the weight vectors are defined as the coordinates of neurons on a two-dimensional space. The neurons are distributed in an "output space" and identified by their positions $\mathbf{w}_i$ which are called *mapping vector* in the model. The change of the vector values means the movement of neurons. Each two-dimensional vector $\mathbf{x}_j$ $(j = 1, 2, ..., m)$ in an input space corresponds to the geographic location of a node j in a telecommunication network, m is the number of nodes.

Given a set of neurons $\mathbf{M} = \{1, 2, ..., i, ..., n\}$. A somatotopical mapping $\phi_\mathbf{r}$ for a traffic routing $\mathbf{r}$ is defined by the mapping from the communication network in the input space $\mathbf{E}$ onto the neural net of output space $\mathbf{A}$:

$$\phi_\mathbf{r} : \mathbf{E} \longmapsto \mathbf{A}, \qquad \mathbf{x} \in \mathbf{E} \longmapsto e \in \mathbf{A}, \tag{1}$$

where e is determined on the following condition:

$$||\mathbf{x} - \mathbf{w}_e|| = \min_{i \in \mathbf{A}} ||\mathbf{x} - \mathbf{w}_i||. \tag{2}$$

Here $\mathbf{w}_e$ and $\mathbf{w}_i$ are the mapping vectors that determine the positions of neuron e and i, respectively. Fig. 1 shows the input and output space of the routing model.

During the learning process the neurons are excited one after another through stimuli of the input space. The groups of the excited

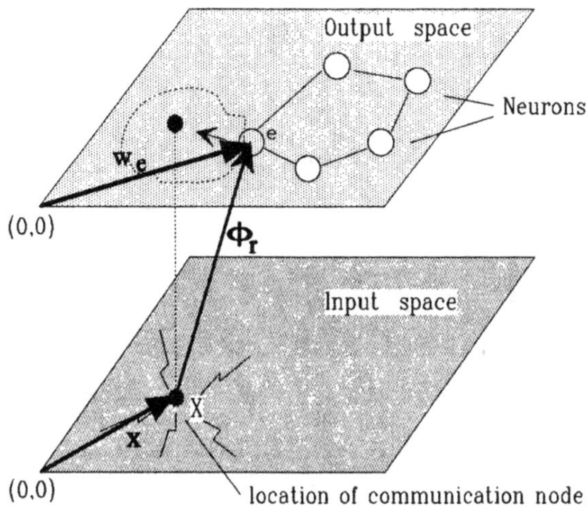

Figure 1: Neural Traffic Routing Model

neurons are determined through the connections to a neuron e which is just excited. This group is defined as connection zone Z_e.

The range of cooperation will be shrunk as iteration time t increases. The initial range of the connection zone $Z_e(s)_{t=0}$ is set to be large, it covers all neurons. The depth s decreases in its value as time goes on, thus localizing the range of maximum activity. At the convergence, the zone reaches to $Z_e(1)_{t=t_{max}} = \{e\}$.

The adaption of mapping vector $\mathbf{w}_i$ corresponding to the input $\mathbf{x}$ is then given by

$$\mathbf{w}_i^{(t+1)} = \begin{cases} \mathbf{w}_i^{(t)} + \epsilon \cdot H \cdot (\mathbf{x} - \mathbf{w}_i^{(t)}) & \forall i \in Z_e(s(t)) \\ \mathbf{w}_i^{(t)} & \forall i \notin Z_e(s(t)) \end{cases}, \tag{3}$$

$\mathbf{x}$ is the two-dimensional vector of the input space, ϵ scales the overall size and H the learning rate. Equation (3) has the desired property of confining any adaptations to the connected neurons in the $Z_e(s)$ and improving subsequent responses to $\mathbf{x}$ there.

For each learning step an input $\mathbf{x}$ is chosen randomly according to the probability distribution density $f(\mathbf{x})$. A random vector $\mathbf{x}$

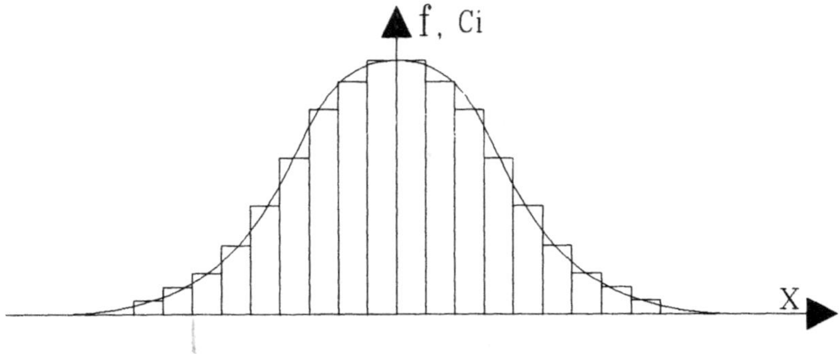

Figure 2: Probability Distribution Density of the Random Number Generator

represents a location of the node in the telecommunication network. A Gaussian distribution is chosen centering on the locations which have large capacities of the communication system. The capacity C_i of each node i of the communication system is ordered by the value. In the location i with large C_i there are higher probabilities to be "stimulated" than in the location with low capacity (see Fig. 2).

After learning process each communication multilink path has the minimal geographic distance subject to the capacity constraints. In the most general case, the metrics on the communication paths could be computed as a function of the bandwidth, average traffic, communication cost, mean queue length, measured delay, and other factors [7]. By changing the weighting $\mathbf{w}_i$ the approach can calculate the "shortest" path according to any one of a number of criteria, e.g. in broadband ATM networks, to minimize the average packet delay subject to the trunk capacity constraints [8][9].

3 Experiments

The algorithm is programmed in *C* language. Computer simulations have shown that an optimal solution can be obtained by means of the neural routing model. Here an example with 38 traffic nodes is given. The example has 7 origins and 7 destinations, respectively. Each path is chosen as a chain of 8 "neurons" in the form of a line. The initial

Table 1: Capacities of Nodes

nodes	1	2	3	4	5	6	7	8	9	10	11	12
C_i	34	15	10	41	20	25	33	22	5	46	39	60
nodes	13	14	15	16	17	18	19	20	21	22	23	24
C_i	18	24	80	46	23	15	85	5	15	19	10	20

values of vectors $\mathbf{w}_i$ are assigned to a straight line between a origin and a destination (Fig. 3(a)). The vectors $\mathbf{w}_i$ are changed iteratively according to equation (3).

The total number of neurons must be at least equal to the number of the nodes. In this example, all lines have together 56 neurons.

Both terminations of the lines are fixed on the locations of the origin node and destination node. Stimuli frequency of each node depends on the given capacity of the node (see Table 1). During the learning process, each iteration causes a local deformation of this line. Initially, as long as H is still long-range, each deformation affects rather large line segments. As the range of H gradually decreases, the deformations along the chain become more localized and more detailed in the line. ϵ is taken to be 0.9.

After 1130 learning steps to the convergence phase, the line takes on its final shape, passing through locations of suitable nodes (Fig. 3(b)). As a result, a multilink path for node-to-node traffic with minimal loss is found.

4 Conclusion

The solution described here shows that the Kohonen's neural network can be applied to practical routing problems. The solution is well defined even for large-size problems, where the parallelism of the neural network is most advantageous. One advantage of the approach is that there is no need to express explicitly the routing problems which are sometimes hard to formalize, e.g. the objective functions. As the next research step, the approach will be extended to ATM network for dynamic network reconfiguration and the reassignment of trunk

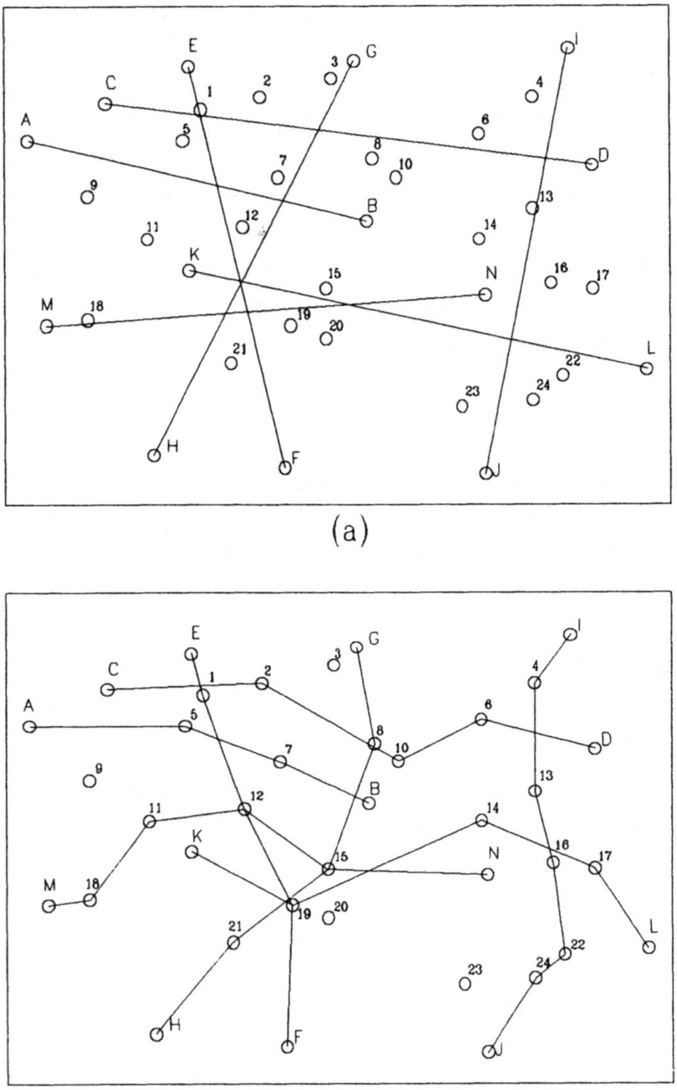

Figure 3: A Routing Experiment of 38 Nodes

capacities in response to traffic changes.

References

[1] J.N. Tsitsklis and D.P. Bertsekas, "Distributed Asynchronous Optimal Routing in Data Networks," *IEEE Trans. Auto. Contr.*, Apr. 1986.

[2] F. Chang and L. Wu, "An Optimal Adaptive Routing Algorithm," *IEEE Trans. Auto. Contr.*, Aug. 1986.

[3] H.E. Rauch and T. Winarske, "Neural Networks for Routing Communication Traffic," *IEEE Control Systems Magazine*, pp. 26-31, April 1988.

[4] T. Kohonen, *Self-Organization and Associative Memory*, 3rd Ed., Berlin: Springer Verlag, 1989.

[5] B. Angeniol, V.G. Croix and J. Le Texier, "Self-Organizing Feature Maps and the Travelling Salesman Problem," *Neural Networks*, vol. 1, pp. 289-293, 1988.

[6] J. Han, I. Wacholder and R.C. Mann, "A Neural Network Algorithm for the Multiple Travelling Salesman Problem," *Biological Cybernetics*, vol. 61, pp. 11-19, 1989.

[7] A.S. Tanenbaum, *Computer Networks*, 2nd Ed., New Jersey: Prentice Hall, 1989.

[8] H. Pan and I.Y. Wang, "The Bandwidth Allocation of ATM Through Genetic Algorithm," *Proc. of IEEE GLOBECOM'91*, pp. 4.4.1-4.4.5, 1991.

[9] X.-Q. Chen and I.M. Leslie, "Neural Adaptive Congestion Control for Broadband ATM Networks," *IEE Proceedings-I*, vol. 139, no. 3, pp. 233-240, June 1992.

Prediction of Access Line Growth

Jerome T. Connor

Bell Communications Research
Morristown, N.J. 07960 USA
connor@faline.bellcore.com

Abstract

The revenue of a telephone company is strongly related to the number of its access lines. Accurate forecasts of a company's future revenue will be possible only if the future access line growth is predicted accurately. This work models the access line growth at the operating company level. A baseline neural network access line predictor is demonstrated that is superior to a simple AR model. Finally, more complicated access line growth models are discussed and access line time series with level shifts are presented.

The revenue of a telephone company is strongly related to the number of its access lines. Accurate forecasts of a company's future revenue will only be possible if the future access line growth can be predicted accurately. One difficulty involved in predicting access line growth of the regional holding companies is due to the limited amount of past data. The regional holding companies have only existed since 1984 and access line growth is generally measured on a year by year basis.

This work models the access line growth at the operating company level. If necessary, the access line growth of the regional holding companies can be determined by adding up the access line growth of the operating companies it owns. This approach has three advantages. First, there exist over one hundred local operating companies compared with only seven regional operating companies. This leads to a greater pool of data from which to build statistical models. The second advantage is the availability of statistics on line growth prior to divesture. Most of the operating companies have been around for a long time and only the management structure above them were affected by divesture. Thus, access line models can be based on a larger time period. Lastly, we expect to find that many of the operating companies have similar access line growth and thus do not need to be modeled individually but can be modeled as a group.

Forecasting of the growth of special services has been investigated [4]. Special services is a generic term referring to services which require special treatment such as private lines or WATS. These special services are often subsets of the total access line count, they are of interest partly due to the greater profit associated with these services. These services were forecasted [4] at the wire center level. The wire center level is more local than the state level investigated here, often there are 500-800 wire centers for a large state.

Often the problem of interest will be how fast will one technology replace another [2] [3]. This problem is often encountered when one type of access line is replacing another. There are several important technological substitution problems currently being encountered in the access line community, such as "How much will cellular replace wired access lines?" or "How quickly will fiber enter the local loop?" Technological forecasting is somewhat difficult due to its non-stationary nature.

One example of how a regional telephone company would use access line predictions occurs at the wire center level. It has been observed at one of the regional holding companies that the optimal design of a SONNET network can change significantly with a change in the number of access lines in the wire center. Thus, if useful predictions of the number of access lines at the wire center level can be obtained, better SONNET network designs can also be obtained.

This work is composed of three sections. Section 1 begins by describing the access line time series. Section 2 describes a baseline neural network model that has been applied to predicting access lines growth. Finally we conclude by mentioning further experiments that are in progress.

1 Access line growth

There are several types of access line time series that forecasters are interested in. The total number of access lines in use is known as the *inservice* time series. The most important type of access line data is the *gain* series, which represents the increase in the number of access lines in service. Due to accounting procedures, the differenced *inservice* time series does not actually equal the *gain* time series, but they are typically very close. The *gain* time series is not the number of people who added the service, that is known as the *inflow*. The difference between *gain* and *inflow* is due to the number of access lines that are discontinued, known as the *outflow*. Due once again to accounting procedures, the *gain* series is usually not equal to the *inflow*

- *outflow* , but typically they are close.

Forecastors are generally interested in the monthly and yearly time series. One interesting aspect of the monthly time series is the seasonality that sometimes occurs. This seasonality is presumed due to such events as schools opening and closing.

Access line growth is not stationary. In Fig. 1, an inservice access line growth time series is shown, the axis have not been labeled for confidentiality In general, the number of access lines grows a few percentage each year. Usually the growth is positive, but sometimes the growth is negative. This can usually be explained by the selling of a part of the business to another operating company or a very bad economic condition in the local economy. The number of access lines in 1991 is actually less than in 1990. Because of the lack of certainty of the negative access line growth, it is wise to use robust modeling techniques that are not overly affected by outliers due to the selling off of portions of the company. Of course, if it is know apriori whether the negative growth is due to a selling of a known amount of company access lines , the differenced time series points that are affected by the selloff can be adjusted to remove the affects of the selloff.

In order to make useful predictions with a neural network, the access line growth, L_t, must be described as a stationary series. The diverging nature of access line growth is removed from the time series by differencing the data. The differenced time series was further normalized in order to compare time series from different operating companies,

$$PL_t = 100\frac{L_t - L_{t-1}}{L_{t-1}} \tag{1}$$

PL_t represents the annual percentage increase in line growth, Fig. 2 displays the annual percentage increase in line growth of a telephone operating company. Once again the axis are not labeled for reasons of confidentiality. As Fig. 2 shows, there are trends within the time series that a modeler would hope to take advantage of.

2 Neural network prediction of access line growth

The annual proportional increase in line growth will be modeled as a nonlinear autoregressive process, NAR(p),

$$PL_t = g(PL_{t-1}, PL_{t-2},, PL_{t-p}) + e_t \tag{2}$$

where e_t is not directly observed additive noise and $g()$ is a nonlinear function which is modeled by a neural network

$$\hat{g}(PL_{t-1}, PL_{t-2},, PL_{t-p}) = \sum_{i=1}^{h} W_i f(\sum_{j=1}^{p} w_{ij} PL_{t-j}) \tag{3}$$

where $f()$ is a sigmoidal nonlinearity and the parameter vectors $\mathbf{W}$ and $\mathbf{w}$ are determined from the data via a least mean square optimization algorithm.

A data set was compiled of the number of access lines of operating companies at the state level on an annual basis from 1983-1991. Many neural networks of various number of hidden neurons and order p were estimated on 500 data points from 1986 - 1989. Backtesting on a cross-validation set from 1984-1985, was used to select the number of hidden nodes along with the stopping point in the training algorithm. The resulting five hidden, $p = 1$, neural network predictor's performance is shown along with a corresponding linear predictor is shown in Table 1. It should be noted that [4] also estimated an AR model of order $p = 1$ in the similar problem of forecasting at the wire center level. The strong performance of the NAR(1) predictor with respect to the AR(1) predictor indicates a nonlinearity at least for $p = 1$.

Table 1: Access Line Predictions

	M.S.E. Training Set 1986-1989	M.S.E. Cross-Val. Set 1984-1985	M.S.E. Testing Set 1990-1991
NAR(1)	103.500	14.517	22.565
AR(1)	141.048	20.356	40.014

3 Conclusions and Future Work

Neural networks were shown to perform favorably vis-a-vis linear predictors for access line predictions. This is only the baseline result however. Further analysis, to be presented at the workshop, of the similarity of access line growth between groups of operating companies will lead to further opportunities for a better segmenting the problem.

Multivariate models with economic variables as input will be investigated. It is very likely that operating companies who are in the midst of a local economic downturn would not be greatly expanding their infrastructure. If this is true, more accurate predictions can be obtained by including local GNP growth as an input into the neural network.

The access line growth time series are affected by outliers. Operating companies will sometimes sell off a portion of their infrastructure to another operating company. This type of access line growth is not predictable from past access line growth and will distort the model.

It should be of interest to builders of access line models due to the impact of technological forecasting on access line time series. It seems reasonable to expect level shifts to be caused by the technological forecasting. Even if the exact level shift is not known exactly, one can infer the value of the level shift and obtain better models, see for example [1]. It is interesting to note that the best technological forecasting models are nonlinear, see for example [3].

Often the cause of level shifts in acces line data are more mundane, such as changes in accounting. One example of a access line data with a level shift is shown in Figure. 3, the name, time, and scale of this data is reserved for buisness reasons.

Finally, further increasing of the data set will be investigated by examining the number of telephones supplied by the operating company. This is historically similar to the number of access lines and has the advantage of going back much further in time.

References

[1] Connor, J. T. "Time Series and Neural Network Modeling," Thesis, University of Washington, 1993.

[2] Keramidas, E. M. and Lee, J.C., "Forecasting Technological Substitutions With Short Time Series," Technical Memorandum, TM-ARH-010835, 1987.

[3] Lee, J.C. and Lu, K. W., "Technological Forecasting With Nonlinear Models," Technical Memorandum, TM-ARH-013499, 1989.

[4] Lee, J.C., Nabil, N. A., and Horng, C, "Forecasting Circuit Counts For Special Services," Technical Memorandum, TM-ARH-015317, 1989.

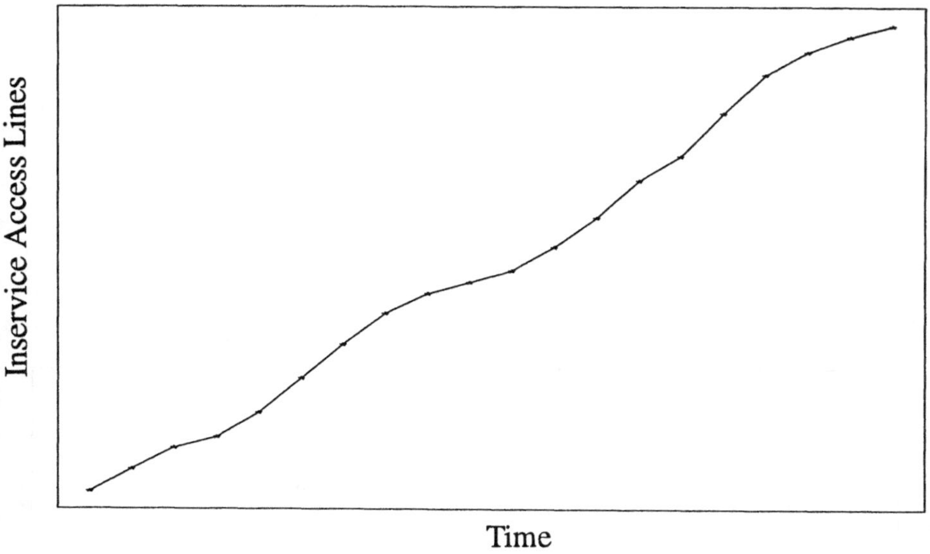

Figure 1: Inservice access lines vs time.

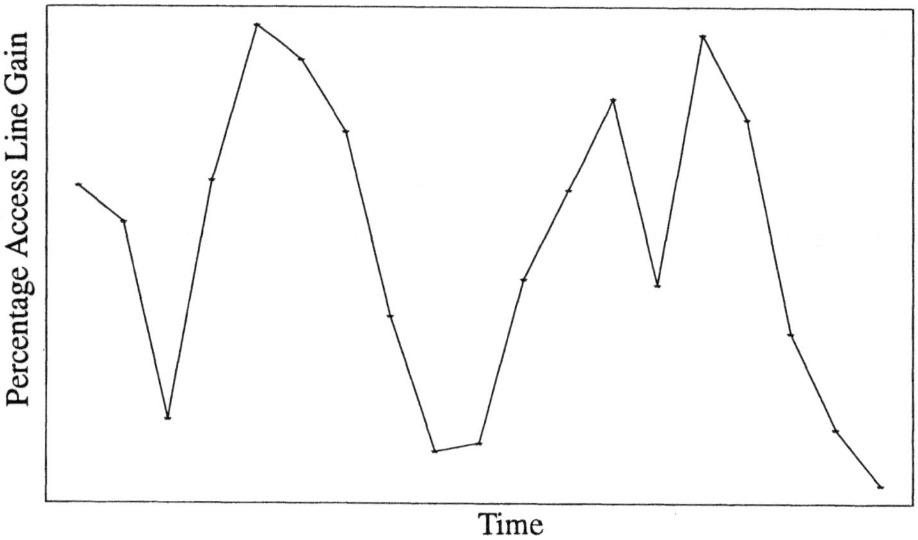

Figure 2: Access line percentage gains vs time.

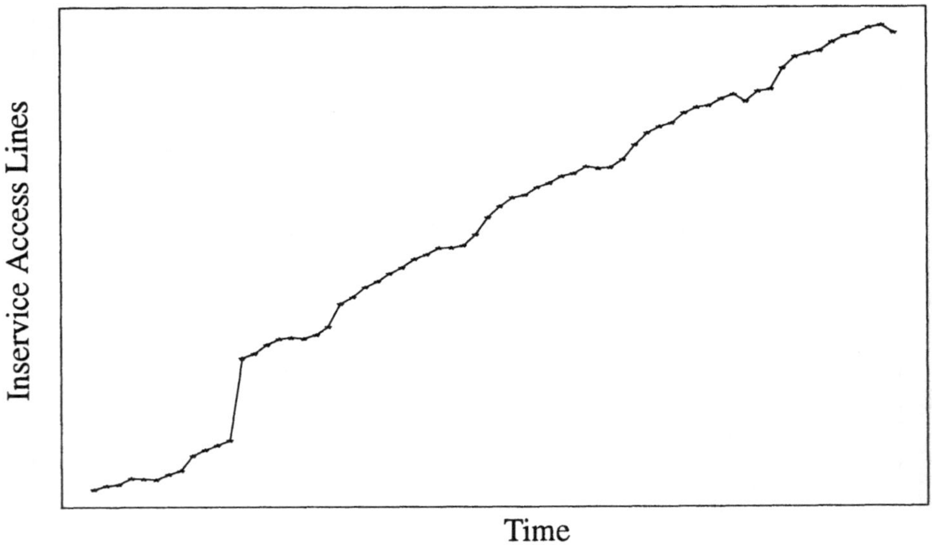

Figure 3: Inservice access lines vs time, for a time series with a level shift.

Toll-Fraud Detection

Ben P. Yuhas

Bell Communications Research
445 South Street, Morristown, NJ 07960
yuhas@faline.bellcore.com

Abstract

Total losses due to telephone fraud are estimated to be as high as billion dollars a year. While a significant amount of work has been done to develop secure interfaces, few of these interfaces are actually deployed. The less sophisticated methods of security that are in place are often rendered ineffective by technical means or by human error. This work assumes that there will always be some unauthorized users and that it is best to supplement a secure interface with the capability of detecting the unauthorized user. This paper describes ongoing research that attempts to use neural networks to detect unauthorized calling-card activity. The results are based upon analysis of actual calling-card activity obtained from the public switched networks

1 Introduction

Total losses due to telephone fraud are estimated to be as high as billion dollars a year. One way to combat fraud is to build more secure systems by placing sophisticated recognition devices at their entrances. Toward this end, neural networks have proven to be effective in automatic face recognition, finger print identification and speaker verification. And yet despite the progress in these areas, such secure interfaces are rarely deployed in the field because of their implementation cost and because of the fear that an increase in the complexity of the interface will result in less usage and lost revenue.

The basic premise of this work is that it is impractical to expect that the public switched network will ever be completely secure. Moreover, most reasonable prevention measures can often be rendered impotent through technical means or human error. As a result there will always be some unauthorized users on the system. Consequently, secured access and restricted usage needs to be supplemented with effective methods of detecting these intruders.

There are many types of fraud that impact telecommunications service providers, including calling-card, PBX/Centrex, subscription and cellular. While each of these results in lost revenue for the service providers, in this paper the focus will be on *calling card fraud*: the use of an illegally obtained Calling Card Account Number (CCAN) to charge phone service. The CCAN and the associated Personal Identification Number (PIN) can be obtained through a variety of methods. Regardless of how it is obtained, once the CCAN and PIN are known, they can be sold to others or used to provide a call setup to an individual for a flat fee (usually below the legitimate rate). The cost of calling-card fraud alone has been estimated at over two billion dollars.

2 Calling Card Usage

When a calling card is used by dialing 0+ the number, you are connected to the pre-subscribed long distance carrier assigned to the telephone being used. Upon receiving the bong tone, you then enter your 14-digit calling card number. The switch identifies the issuer of this calling card and directs the query to the appropriate data base.

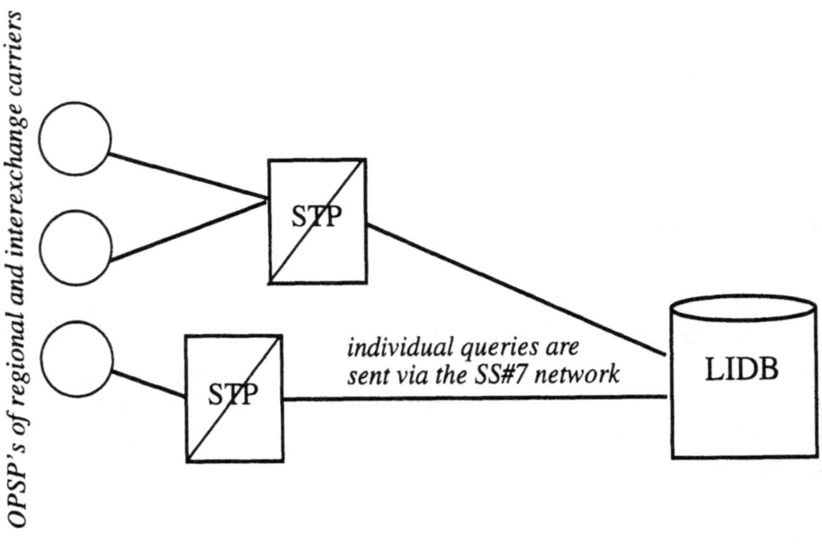

Each calling-card number (CCN) and the associated Personal Identification Number (PIN) are validated against the data base maintained by the issuer of that card. So, for example, AT&T, MCI and Local Exchange Carriers (LEC) each maintain their own data bases. For the purposes of this paper, we will focus on the LECs calling cards which are validated against the LECs' data bases, called the Line Information Data Base (LIDB). This database is also queried for bill-to-third and collect calls. The individual queries to LIDB are primarily transmitted via the SS#7 messaging network. A query will include, among other things, the originating and terminating numbers, the billing number, the time of day and the date. The LIDB's response to the query is returned to the switch, which then proceeds with the processing of the phone call.

LIDB currently counts successful CCN queries. Abnormal, and thus suspicious, usage is defined as those calling-card calls exceeding a given threshold within a specified amount of time. Two levels of thresholds are applied: a high-usage threshold (HUT) and an automatic deactivation threshold (ADT). When thresholds are exceeded fraud messages (FM) are sent out to fraud investigators in the following order: a High Usage Threshold Exceeded (HUTE) message followed by Incremental High Usage Threshold Exceeded (IHUTE) messages and finally an Automatic Deactivation Threshold Exceeded (ADTE) message. In addition, for each individual query sent to LIDB, a detail record is generated and saved. The analysis that will be reported on in this paper is based on over 10 million of these detail records that have been obtained from in-service LIDBs.

3 Enhancing Calling-Card Fraud Detection

This research aims to improve fraud detection by first improving the use of those systems in place and secondly, by defining more advanced automatic fraud detection techniques which use more of the available information. The goal of this data analysis is to move beyond anecdotal methods of identifying fraud based upon hypothetical scenarios and attempt to identify methods of fraud detection based on observed patterns and measurable quantities.

To go beyond the threshold method of detecting fraud by defining more sophisticated detection techniques, we need to look at broader patterns of usage, such as when and where a customer usually makes his calls from, as well as the normal destination of these calls.

The original intent of the work was to assemble examples of fraudulent usage and to train a network to distinguish these from non-fraudulent usage. As is often the case, obtaining labeled data proved to be difficult. In our case, the problem was made worse by organizational structures in the regions that often separate the fraud investigators from the final resolution of the bill. Consequently, a fraud investigator may identify what is perceived to be a potential case of fraud, but may never have it confirmed as fraud. The exception is when the fraud investigators are able to contact the calling-card holder, which is always attempted but usually not successful.

Failing to have sufficient training data forced us to consider non-supervised methods such as clustering. The new task would be to cluster the users based on calling characteristics and to see if the few identified fraud cases end up in the same cluster. If such a cluster existed, then it would merit investigating the other members of that cluster. Prior to clustering the data, a representation of the card-holder's profile needs to be built.

4 The User's Profile

The individual queries associated with a given calling card are not very useful. It is only by creating an aggregate picture that it is possible to distinguish normal from abnormal usage. After accumulating all of the detail associated with a given user it is next necessary to summarize this detail.

For our initial experiments, a vector containing nine variables was defined, which included:

- total number of queries
- fraction overseas calls
- fraction calling card vs. non-calling card calls, such as bill-to-third
- fraction day vs. night
- number of different originating area codes/ number of unknown originations
- number of different terminating area codes/ number of unknown terminations
- smallest interval of time between arriving queries (interarrival time)

The dynamic ranges of these nine variables vary greatly, with the three fractional values restrained between 0 and 1, while the total number of queries may approach 1000.

5 Clustering

Given these users' profiles we would like to find a means of identifying those variables that identify the user as fraudulent. One way to accomplish this is to find the dimension on which known cases of fraud are similar. Clustering algorithms attempt to find natural groupings in the data. The algorithm used here is based on a version by Yoshiro Miyata made available by Andreas Stolcke via anonymous ftp at **`icsi-ftp.berkeley.edu`**. The algorithm starts by treating each vector as a group and then finds the closest groups and merges them. This process is repeated until all vectors belong to a single group. While any reasonable metric could be used, the default is Euclidean distance. For a more complete description of clustering see Duda and Hart (1973).

6 Experimental Results

Using three days of data obtained from a single mated-pair of LIDB data bases, over 340,000 distinct users were seen. Clustering on this many vectors proved to be impractical and a subset of the largest 500 users was used instead. In this subset, there were 20 cases of identified fraud.

The final preprocessing of the data that remained prior to clustering was normalization. Since the Euclidean distance was used to measure the distance between vectors, the effect of normalization is profound. In particular, the dynamic range of the individual components will directly effect how much importance they are given when measuring relative distance between vectors. Put another way, variables with larger values will tend to have a greater impact on the distance between two vectors.

In the first experiment, all components of the vectors were normalized so that their values ranged between 0 and 1. This forced the algorithm to treat all nine dimensions with equal importance and reduced the presence of explicit bias in the representation. While this was intuitively attractive, it proved to be unsuccessful. Using this normalization, the individual cases of fraud were widely distributed across multiple clusters. It wasn't until the clusters grew to include almost a quarter of the data that a cluster would include four or more cases of fraud. Consequently, these clusters provided little help in identifying suspicious subsets of the data.

In the second experiment, the components were left in their original form. This resulted in some dimensions having dynamic ranges 2 and 3 orders of magnitude larger than others. As a result, the total query count and the query interarrival time are emphasized. This proved to be very successful, with 7 of the 20 cases of fraud landing in a single cluster of 16. By further observing the vectors within this cluster of 16 users, we find that the magnitude of the of the total query counts are close together and the interarrival time tends to be small. The magnitude of these total query counts could be used to refine the thresholds within LIDB.

7 Conclusions

This paper presents some early results from an analysis of LIDB queries for the purpose of identifying calling-card fraud. The lack of a sufficient amount of labeled data made it necessary to pursue non-supervised techniques, such as clustering. In these preliminary experiments, we find that there are dimensions on which a users' calling habits can be clustered to help identify cases of fraud. At this point, the clustering appears to be relying on a very small subset of an individual calling-card's profile. It is hoped that by investigating alternative metrics for the clustering algorithm and exploring the impact of various normalizations that more of the available information will become useful. In the near term, these clusters may help us identify better thresholds for use in LIDB. Better thresholds can reduce the number of false alarms associated with setting the thresholds too low and increase the sensitivity to fraud in those cases where the thresholds were set too high.

References

[1] Duda, R.O. and Hart P.E., *Pattern Classification and Scene Analysis*, John Wiley & Sons, Inc., New York, NY, 1973.

Multistage Information Filtering Using Cascaded Neural Networks

Thomas John

Southwestern Bell Technology Resources
St. Louis, MO. 63141
john@sbctri.sbc.com

Abstract

This paper describes a multistage neural network classifier for filtering information. Classifiers based on adaptive mapping networks are used to incrementally refine information as it passes through a cascade. Results are derived showing that such a refinement process can be performed. Experimental results are derived to determine the computational efficiency and accuracy of the filter cascade. Applications to information distribution and information bases are suggested.

1 Introduction

Although there has been much work on creating information retrieval systems and information filters such as Wais [1], there has been little work on arranging these filters as cascades which refine information incrementally. We propose an arrangement of neural networks which classify information in stages so that the final recipients get the most targeted information while the intermediate hubs produce less focussed output. These neural networks are trained over a period of time to recognize the information that is needed at each node of a distribution network. The neural networks which perform the classification have separation properties which enable them to learn to differentiate between likes and dislikes.

We will use the example of a library to illustrate how the cascaded classification method works. Suppose that all information at the level of a general distribution facility is classified into various broad subject categories.

Suppose that a central library distributes the information based only on particular subject interests. The branch libraries can redistribute the information based on knowledge about the interests of individuals. The library model works well though it is slow. With large amounts of information, for example as in the electronic superhighway, a lot of this classification and association mechanism should be automated.

We describe a neural mapping network where this automatic refinement property can be obtained. As in Adaptive Resonance networks, granularity of classification of our mapping networks can be controlled by adjusting an attentional vigilance parameter [2]. We use a mapping network which is different from the standard Artmap scheme in [3]. We describe a classification and association scheme which incorporates a learning and refinement procedure.

2 Multistage mapping network

Neural networks can be used for classifying information. ART 1 is an unsupervised classification network. On a given collection of inputs, an ART 1 network stabilizes by learning subsets of these inputs, for details see [4]. Subsequently any of the inputs match one of these subsets. The number and size of subsets vary depending on an attentional vigilance parameter. In general, a higher vigilance produces a larger number of subsets.

ART 1 networks we consider consist of two layers F_1 and F_2. Input to the net is presented at layer F_1, these inputs are propagated to the layer F_2 via bottom-up links, activating the neurons in F_2. On the other hand, F_2 will send a top-down expectation to F_1, a match between the bottom-up and top-down patterns results in resonance, i.e. recognition. We skip details on the ART 1 model, see [4], except to note that fineness of classification is controlled by an attentional vigilance parameter v. Suppose an ART 1 network has been trained sufficiently to classify all the information it receives into several classes. Suppose further that for any one class, all items in that class are related pieces of information. A filter can be constructed by keeping only those classes which contain interesting items and by discarding the information in the rest of the classes.

Consider a cascade of filters $\Phi_1, \ldots, \Phi_n$ which progressively refine information, i.e. Φ_1 will filter out very little while Φ_n will filter out most of the information. One way to create such a cascade is to use n ART 1 classifiers, all of them identical except for the vigilance parameters $v_1, \ldots, v_n$. To achieve progressive refinement, the vigilance parameters will progressively increase, i.e. $(\forall i < n) v_i < v_{i+1}$. We will show how the filters $\Phi_1, \ldots, \Phi_n$ can be built so that each filter refines information as needed. Thus generality will be preserved to the extent necessary, while targeted information can be delivered where required. A few details about ART 1 is needed for the results which will be derived.

We will derive results about subsets of ω, the set of natural numbers (i.e. positive integers.) Things learned by ART 1 network can be considered as sets of natural numbers by considering F_1 to be made up of neurons numbered $1, \ldots, n$. A stable F_2 neuron therefore can be represented by a set $c \subset \omega$, which we will call a *code*. If a set $I \subset \omega$ matches the code c, we write this is $I \sim c$. We can also consider a *class* of sets C which consist of those subsets of ω which match a particular code c, i.e. $C = \{I \subset \omega \mid I \sim c\}$. Here, an input I, $I \sim c$ if $\mid I \bigcap c \mid / \mid c \mid$ is large enough in an ordering involving all $c \in F_2$ and the vigilance criterion $\mid I \bigcap c \mid / \mid I \mid > v$ is met.

ART 1 networks have a property which is suitable for making cascading filters. To make this explicit we consider a notion of distance between sets. For $I, J \subset \omega$, a distance between I and J can be defined as a normalized symmetric difference, $\delta(I,J) = \mid I - J \mid / \mid I \mid + \mid J - I \mid / \mid J \mid$. The distance $\delta(I,J)$ can be considered as a measure of how much two sets I and J differ, note that $\delta(I,I) = 0$. A property of ART 1 networks is that the distance between sets in the same class decrease as vigilance increases. We will state this in terms of the *diameter* of a class C which is defined as the maximum of $\delta(I,J)$ for all $I, J \in C$.

Theorem 1 *Suppose N_v is an ART 1 network which has been trained with vigilance v until it has stabilized on all inputs. Then for any code $c \in F_2$ and for the associated class C, the diameter of C is bounded by a term which decreases as the vigilance parameter v increases.*

Proof:

Let I, J be in C. By definition I and J match $c \in F_2$. By the direct access theorem (Theorem 6 [4]), $c \subseteq I$ and $c \subseteq J$. By the vigilance criterion, $\mid I \bigcap c \mid / \mid I \mid > v$ i.e. $\mid c \mid / \mid I \mid > v$. Now $\mid c \mid + \mid I - c \mid / \mid I \mid = 1$ so that $\mid c \mid / \mid I \mid = 1 - \mid I - c \mid / \mid I \mid > v$. Therefore $\mid I - c \mid / \mid I \mid < 1 - v$. Since $I - J = I - I \bigcap J$ and $I \bigcap J \supseteq c$, $\mid I - J \mid \leq \mid I - c \mid$ hence $\mid I - J \mid / \mid I \mid \leq \mid I - c \mid / \mid I \mid < 1 - v$. Similarly $\mid J - I \mid / \mid I \mid < 1 - v$. Therefore $\delta(I,J) < 2(1-v)$. The term $2(1-v)$ decreases as v increases. □.

Theorem 1 can be used to refine filters. Suppose a filter Φ is created by classifying all information using an ART 1 network N_v and subsequently by associating a function f with the codes such that $f(I) = 1$ iff I matches a code c such that an item J which matched c was previously considered to be interesting. Interests are complex notions, but we can consider what is interesting to be defined by an analytic relation. Analytical, or Σ_1^1 relations, are projections of Borel sets [5]. Very general concepts can be described using analytical sets, for example faithfulness, guarantee and other temporal properties are trivially expressed as analytic predicates [6]). The network filter Φ consisting of the network N_v and the function f can be considered as a *mapping network* M 1 which has been previously considered in [7]. To avoid complex notation, let us consider an analytic set A to be "possible worlds" where there is an "enduring" interest in pieces of information $I \subset \omega$. An ART 1 network N_v can approximate A by constructing certain codes c such that for any I of interest according to A, there is a class C and item $J \in C$ such that $\delta(I, J)$ is small. Meanwhile, if $B \in \Sigma_1^1$ expresses what is not interesting, then codes associated with B should be distinct from the codes associated with A. A classical result, ascribed to Blackwell in the context of probability theory [8], can be extended to provide such a separation result.

Theorem 2 *Suppose that A and B are Σ_1^1 sets such that $A \bigcap B = \emptyset$. Then there is a mapping network which can approximate a Borel set C to arbitrary degree of accuracy, where C has the property that $A \subseteq C$ and $B \bigcap C = \emptyset$.*

Proof(sketch):

A formal proof can be given using infinite games [5], which establishes the relationship between a Borel set C as above and a mapping network which we construct. Here we will describe the construction of the mapping network. Suppose a network N_v does not distinguish between the interests in A and the uninteresting things in B. Then there is a code c such that $(\exists)(I, J \in C)$ where I is interesting and J is not. Then we can refine N_v to a new network N_v' by increasing vigilance to $v' > v$ and by running sets like I and J through N_v' until new codes are formed such that I and J are in separate classes. If any I matches such a code c in the future, then the distance between I and any other I' in the same class C is bounded by the formula derived in Theorem 1. The vigilance v' can be selected to provide a desired degree of accuracy. □.

Theorem 2 can be extended to consider the construction of mapping networks all over a network of information filters. A network does not have to be a cascade. Filter networks, consisting of embedded cascades, form a natural model for information flows.

Theorem 3 *If $A, B \in \Sigma_1^1$ are disjoint and represent concepts to to be separated by a classifier N_v , and $A', B' \in \Sigma_1^1$ are disjoint and represent concepts to be separated by a classifier N_u , then N_v and N_u can be refined to simultaneously approximate Borel separators C and C' such that*

$$A \subseteq C, B \bigcap C = \emptyset, A' \subseteq C', B' \bigcap C' = \emptyset$$

Proof: Again, we will only sketch a construction of N_v and N_u. If the classification of N_u is not consistent with the classification of N_v then N_u can be refined until the accuracy is at a desired level. Similarly N_v can be refined according to the needs of N_u. This process stabilizes once the desired level of accuracy is achieved in N_v and N_u. □

Theorem 3 states in particular that it is possible to incrementally refine information based on pooled interests. If A_1 and A_2 represent interests of two different individuals, then $A1 \bigcup A_2$ represents their joint interests. If several such A_i are pooled together at a central site, then the fineness of classification at the central site may be less than the fineness required for each individual A_i. The required level of accuracy can be adjusted using the construction given in the theorems. This construction therefore tells us how to achieve the "library model" of classification mentioned in the introduction.

3 Experimental Results

The results in the previous section show that it is possible to create a cascade of incrementally refining filters by adjusting attentional vigilance parameters in ART 1 classifiers. In this section we indicate the computational demands of doing this type of incremental refinement.

ART 1 networks are unsupervised classifiers. This means that they learn most of their codes before anyone has started indicating things of interest. To ensure that most items of interest match some learned code, the networks have to stabilize on a reasonably large collection of inputs. This requirement is computationally demanding, and training runs have required "forty days and forty nights" on conventional Sun Sparc class machines. The training problem however is ideally suited for massively parallel computers. On a MasPar MP-1 with 4K processors, a network presented with about 10K samples stabilized in approximately five hours. This indicates that training samples as large as 30K can be used to train a filter overnight on a similarly equipped massively parallel processor.

Once codes are learned, classification can be done much faster, in near realtime on conventional workstations. Classification speed depends on the number of codes. One advantage of a cascade of filters $\Phi_0, \ldots, \Phi_n$ is that relatively crude classification can be done by Φ_0 while highly targeted individual classification can be done by Φ_n. A crude classifier will operate at a lower level of vigilance, and will have to deal with a fewer number of codes. Table 1 summarizes shows the number of codes formed on approximately 10K samples at various levels of vigilance.

We also describe a set of experiments over a reasonably large data set showing how fast the network cascades seem to stabilize. According to the theorems in the previous section, it is possible to refine a network to differentiate between likes and dislikes, but it is difficult to simulate a human expressing likes and dislikes. Instead, we considered the existing lack of

Table 1: Number of codes at various vigilances

Num Trials	*Vigilance*	*Num Codes*
20	0.1	178
30	0.2	632
30	0.3	1142
30	0.4	1548
30	0.5	1954
30	0.6	2603
40	0.7	3097
40	0.8	3525
50	0.9	3945

Table 2: Consistency between cascade layers

v_1	v_2	*Ratio*
0.2	0.9	0.92
0.2	0.7	0.82
0.2	0.4	0.57
0.4	0.7	0.79
0.4	0.9	0.92
0.7	0.9	0.92

consistency between classifications at various levels of vigilance. If the filters Φ_i are trained at various levels of vigilance then they have to be retrained using Theorem 2 only to the extent that the classification at a low vigilance v_1 is inconsistent with the classification at a higher vigilance v_2. We can describe this consistency as the proportion of classes C_2 in N_{v_2} such that $(\exists C_1 \in N_{v_1})(\forall I \in C_2) I \in C_1$. For any such class C_1 we can assume that C_1 naturally splits into many classes C_2 as vigilance is increased from v_1 to v_2. The experiments were done on approximately 10K samples not used during the training phase.

Table 2 indicates that the ratio in the last column is constant when v_2 is high, regardless of v_1. This suggests that the accuracy required at the later filters Φ_n may not significantly affect the classification at the earlier filters Φ_i, thereby suggesting that the amount of retraining needed in a cascade to get all the levels consistent may not be excessive.

4 Conclusions

We have described a multistage filtering method where filtering progressively improves in accuracy. We have shown that it is possible to correct the filters to reflect interests at various levels of accuracy. We have also described some experiments which suggest that multistage filtering is computationally feasible, especially using massive parallelism.

Due to the unsupervised nature of classification described here, it is not necessary to perform a separate classification for each user. In fact, by controlling vigilance properly it is possible to classify for pools of users at the hubs of the networks. Multistage filtering can also be used for creating information bases which can be efficiently searched due to the fact that classification is done prior to searching.

References

[1] C. Stanfill, R. Thau and D. Waltz, "A parallel indexed algorithm for information retrieval," *Proc. 12th SIGIR Conference*, 1989, pp. 88–97.

[2] G. Carpenter and S. Grossberg, "Neural dynamics of category learning and recognition", *The Adaptive Brain* I, New York: North-Holland, 1987, pp. 238–286.

[3] G. Carpenter, S. Grossberg and J. Reynolds, "ARTMAP: Supervised real-time learning and classification of nonstationary data by a self-organizing neural network," *IEEE Expert*, vol. 6, 1991.

[4] G. Carpenter and S. Grossberg, "A massively parallel architecture for a self-organizing neural pattern recognition machine", *Computer Vision, Graphics and Image Processing* vol. 37, 1987, pp. 54–115.

[5] Y. Moschovakis, *Descriptive Set Theory*, New York: North-Holland, 1980.

[6] Z. Manna and A. Pnueli, *A Hierarchy of Temporal Properties*, Stanford University Technical Report, 1987.

[7] T. John, "Relationship of ART 1 models to continuous functions", *BP Research*, Topical Report Number 8094, 1990.

[8] R. Van Wesep, "Separation principles and the axiom of determinateness", *J. Symbolic Logic* vol. 43 (1978) pp. 77–81.

Temporal Difference Learning Applied to Continuous Speech Recognition

Laurens R. Leerink, Marwan A. Jabri

Systems Engineering & Design Automation Laboratory
Department of Electrical Engineering
The University of Sydney
NSW 2006, Australia
laurens@sedal.su.oz.au, marwan@sedal.su.oz.au

Abstract

This paper evaluates the performance of two reinforcement learning algorithms on four, increasingly complex tasks of continuous speech recognition. Simulations reveal that when errors are examined only at word boundaries, both algorithms provide performance comparable with that of a fully supervised system.

1 Introduction

As a first step towards a continuous speech word recognition system trained only by reward and punishment, this paper examines the performance of the temporal difference (TD) algorithm [11] and the infinite input duration (IID) algorithm [7] for four increasingly complex tasks in continuous speech recognition:

- End-of-word detection from phoneme sequences.
- Word recognition from phoneme sequences.
- Phoneme recognition from speech spectra.
- Word recognition from speech spectra.

In the reinforcement learning (RL) paradigm there is an agent performing actions in an unknown (usually Markovian) environment, receiving rewards or feedback which depend arbitrarily on the past history of actions. In the context of speech recognition, the system observes a sequence of temporal inputs (phonemes or spectral data) which culminate in an event at which the system is expected to perform an action (identify the correct word).

A fundamental difference between the speech recognition task and typical "learning control" problems is that the actions of the agent cannot influence the sequence of system states. The agent can only observe the sequence and maximize the reward received (or minimize the punishment) by taking appropriate action at the correct point in time.

There are two intrinsic problems that have to be solved when applying RL to speech recognition; the *temporal credit assignment problem* where the agent has to determine how the punishment and reward received from the environment is assigned to the individual actions taken by the agent; and the *perceptual aliasing problem* when the agent cannot distinguish clearly between different world states.

To evaluate if the RL framework can be applied successfully to speech recognition, in this initial study we have constructed an environment which provides punishment at word boundaries. This eliminates the temporal credit assignment problem as the agent is taught *when* to act. The agent is thus exposed to a supervisor only at word boundaries, and as the outputs are Boolean this is best classified as "partially" supervised learning. However, the perceptual aliasing problem remains and has to be resolved by the algorithms.

This is an intermediate step to a complete RL system in which a single reward is received at the end of a continuously spoken sentence.

2 The Algorithms

The temporal difference (TD) algorithm is one of the most widely used algorithms in reinforcement learning research. Sutton [11] considered speech recognition as one of the suitable applications of the algorithm. The reason provided was that when humans see or hear things they receive a stream of input over time and constantly update their hypothesis about what they are seeing or hearing.

However, to the author's knowledge the TD algorithm has not yet been applied to the speech recognition problem. Tesauro's [13] TD-Gammon program has been reported to be the first application of the algorithm to a complex, non-trivial task.

Two problems have to be overcome when applying the TD algorithm to word recognition. The algorithm requires that the system to be predicted

- be dynamic (theoretically an ergodic Markov process).
- must have a state that can be observed and is evolving in time.

One solution is to construct a Markov chain for every word as is done in conventional statistical speech recognition. We then have an absorbing Markov process (all sentences start and end in silence) with stochastic state transitions. Reinforcement learning can then be used as an alternative to versions of the dynamic programming algorithm.

Our approach has been, that for the network to be useful, it must extract all relevant information from the input stream itself, and learn the construction of words through interaction with the environment. Obvious advantages are that the system does not need to be continually provided with the correctly classified output, the current input and the corresponding state. The problem of providing an evolving state is solved by two methods

- Construction of an adaptive world model [5].
- Using tapped delay lines.

The world model is a recurrent neural network which can be trained by gradient decent using variants of the backpropagation algorithm. The recurrent connections of this network are available to store temporal/contextual information.

When exploited by the training algorithm these recurrent connections can develop a representation which summarizes all relevant information from the past which, when combined with the present inputs, can provide enough information to model the behavior of the system. The activations of these recurrent connections are used as an approximation of relevant information for the word recognition task, and supplements the inputs presented to the neural network trained by the TD algorithm. This combination will be referred to as the TD World algorithm.

The concept of tapped delay lines has been widely used in speech recognition, an overview is provided by [3]. For all tasks a delay line length of three was used.

The IID algorithm does not require the system to be an ergodic Markov process and can be used to train a recurrent neural network directly. Previous applications include playing tick-tack-toe [7], emulating a flip-flop and "non-Markovian" pole balancing [10].

3 Results

Training data was derived from the TIMIT [6] acoustic phonetic speech corpus. For all problems the performance were compared to that of a fully supervised network trained using the backpropagation through time (BTT) algorithm [9]. The architecture used in our experiments is shown below in figure 1. It can be classified as a two-layer partially recurrent network, and is a simplified version of the Elman [2] network which has the hidden layer removed. Robinson and Fallside [8] have named it a recurrent error propagation network and have applied this network to the problem of phoneme recognition.

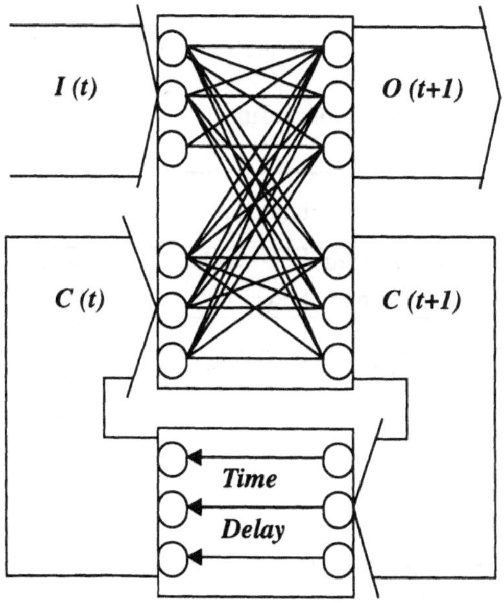

FIGURE 1. The Error Propagation Network

In figure 1 the context units are represented by the *C(t)* neurons. During normal operation the input vector at time *t* are applied to nodes *I(t)*, and during the feed-forward pass values are produced at both the output nodes *O(t+1)* and the context units *C(t+1)*. The values of the context units are then copied back to the input layer *C(t)* for use as input in the following time period.

For recognition from phoneme sequences the network was trained on 50 sentences by different speakers, and tested on the same sentences spoken by another group of different speakers. The training set is 25% of that used by [1] who examined the same problem, but uses a testing set of the same size. The training set was reduced to keep simulation times within practical limits.

For recognition from digitized speech the data was pre-emphasized, a 32ms Hamming window applied and the Fourier transform computed. The input to the neural network was the cube root of the resulting spectral data divided into 20 bark-scaled channels with a separate power input. The networks were trained on 5 pronounciations of the following sentences

This was easy for us.
Jane may earn more money by working hard.

She is thinner than I am.
Bright sunshine shimmers on the ocean.
Nothing is as offensive as innocence.

and tested on the 2 additional versions of the same sentences. All sentences were spoken by different speakers.

The following table summarizes our results :

	Phonemic Input Data		**Spectral Input Data**	
	End-Of-Word Detection	Word Recognition	Phoneme Recognition	Word Recognition
TD-World	-	78%	69%	-
IID Algorithm	64%	83%	73%	61%
Fully Supervised BTT	67%	83%	76%	57%

4 Discussion

The TD World algorithm did not perform well on the end-of-word recognition problem. Possible reasons are

- The TD algorithm used in this manner is best suited to predicting an output evolving in time, whilst this task is to determine a specific moment in time.
- The world model could not predict the next phoneme correctly (44% correct), not providing enough information to solve the perceptual aliasing problem.

Using the tapped delay-line in conjunction with the TD algorithm was not successful. The most likely reason is that the average word contains 27 spectral values, so a delay line of length 3 is is obviously not sufficient. However, when used concurrently with the TD-World and IDD algorithm training times were approximately halved.

5 Conclusions

The simulations reveal the following advantages of the IID algorithm compared to the TD World algorithm:

- No need for a previously (or concurrently) trained 'world' model.
- Training can be accelerated using existing optimization algorithms. Our implementation used the Jacobs [4] Delta-Bar-Delta weight update rule with the entropic cost function [12].

- The algorithm uses recurrent neurons to construct 'credit assignment paths', the standard TD algorithm seems unable to take advantage of the recurrent neurons.
- Training times are approximately an order of magnitude faster (partly due to the optimization methods).

It will be interesting to see if these advantages remain when the system is extended to include full RL (one reward at the end of the sentence).

Research is continuing in this area, with the emphasis on word recognition from spectral data.

References

[1] R.B. Allen and C.A. Kamm, "A Recurrent Neural Network for Word Identification from Continuous Phoneme Strings", In *Advances in Neural Information Processing Systems*, vol. 3, Morgan Kaufmann, 1990.

[2] J.L. Elman, "Finding Structure in Time", *Cognitive Science*, vol. 14, 1990, pp. 179-211.

[3] J.A. Hertz, R.G. Palmer and J.A. Krogh, *Introduction to the theory of neural computation*, Addison-Wesley, 1991.

[4] R.A. Jacobs, "Increased Rates on Convergence Through Learning Rate Adaptation", *Neural Networks*, vol. 1, 1988, pp. 295-307.

[5] M.I. Jordan, "Generic constraints on unspecified target constraints", In *Proceedings of the First International Joint Conference on Neural Networks*, Washington, DC, San Diego, 1989.

[6] L. Lamel, R. Kassel S. and Seneff, "Speech database development: Design and analysis of the acoustic-phonetic corpus", *Proceedings of the DARPA Speech Recognition Workshop*, 1986, pp. 100-109.

[7] A.J. Robinson and F. Fallside, "The utility driven dynamic error propagation network", Technical Report CUED/F-INFENG/TR.1, Cambridge University Electrical Engineering Department, 1987.

[8] A.J. Robinson and F. Fallside, "An error propagation network speech recognition system", *Computer Speech and Language*, vol. 5, 1991, pp. 259-274.

[9] D.E. Rumelhart, G.E. Hinton and R.J. Williams, "Learning Internal Representations by Error Propagation", In *Parallel Distributed Processing*, vol. 1, chap. 8, 1986.

[10] J. Schmidhuber, "Reinforcement Learning in Markovian and Non-Markovian Environments", In *Advances in Neural Information Processing Systems*, vol. 3, Morgan Kaufmann, 1990.

[11] R.S. Sutton, "Learning to Predict by the Methods of Temporal Differences", *Machine Learning*, vol. 3, 1988, pp. 9-44.

[12] S.E. Solla, E. Levin, and M. Fleisher", Accelerated Learning in Layered Neural Networks", *Complex Systems*, vol. 2, 1988, pp. 625-639.

[13] G. Tesauro", Practical Issues in Temporal Difference Learning", *Machine Learning*, vol. 8, 1992, pp. 257-277.

Vector Quantization Using Tree-Structured Self-Organizing Feature Maps

Tzi-Dar Chiueh, Tser-Tzi Tang, and Lian-Gee Chen

Department of Electrical Engineering
National Taiwan University
Taipei, Taiwan 10617
chiueh@bronco.ee.ntu.edu.tw

Abstract

In this paper, we propose a modified SOFM model with a binary-tree cell structure. During and after training, the centroids of the neurons in the model always constitute a binary tree in the higher-dimensional input space. This model is used to design codebooks for vector quantization coding of images. Simulation results show that the acquired codebook is more diversified and has a tree structure. With these two characteristics, it not only produces better-quality images but also expedites the encoding (quantization) process.

1 Introduction

Vector quantization (VQ) has been an efficient method of data compression, especially for speech and images [1, 2]. One of the main reasons for the popularity of VQ is that decoders of VQ-coded data are of minimal complexity, consisting only of a look-up table. Codebook design is an important factor that determines the quality of vector-quantized images. The most famous algorithm for designing a codebook, i.e., the LBG algorithm [3], generates a sub-optimal codebook by iteratively improving on the codebook using an average-distortion (or mean-squared-error, MSE) criterion. The mechanism of iteratively producing better codebooks is *gradient descent*, which implies that the MSE monotonically decreases as the iteration process goes on. The use of gradient descent also suggests that the LBG algorithm may get trapped at local minima. Therefore, the performance of the LBG algorithm depends on the initial condition (codebook) chosen.

Another issue that influences the search of a "good" VQ system is the encoding complexity. For an unstructured codebook, like the one generated by the LBG algorithm, the encoding process requires computing the distortion between the incoming pattern and all the codewords, then finds the codeword with minimum distortion. This procedure usually calls for a time complexity proportional to the number of codewords. Many studies have embarked on circumventing this difficulty by using structured codebooks. The most studied structured VQ is the one that uses codebooks with *binary-tree* structure. Previous tree-structured encoding (search) methods usually impose a binary-tree-structure on an already-designed codebook. Some examples are: hyperplane-based search [4], DCT-based features search [5]. There are also works on building structured codebook from scratch. For example, Riskin and Gray proposed a greedy method to design a codebook with inherent tree structure [6]. However, the obtained codebook is not balanced and hence the encoder complexity may be high.

Self-Organization Feature Maps (SOFM) proposed by Kohonen has been known to be capable of learning *structured* clusters without any supervision. Even without exploiting its final structure, SOFM has been shown to yield VQ codebooks that are better than those generated by the LBG method [7]. Also the SOFM learning is not affected by the choice of initial condition, which can be attributed to the fact that the SOFM learning is similar to *annealing* especially if the learning rate and the neighborhood both decreases exponentially. Therefore, the SOFM algorithm is less likely to be stuck at local minima and it usually yields better codebooks than those designed by gradient-based methods. In this paper, we shall propose a new SOFM with a binary-tree-structure neighborhood, which we call Binary-Tree-structured Self-Organization Feature Maps (BTSOFM). The BTSOFM is then applied to designing tree-structure codebooks for vector quantizer used in image coding. Since a tree-structure is retained throughout the training process, we expect that the designed codebook will not only represent images better but also be more suitable for tree-search encoding.

2 Binary-tree-structured SOFM

In the original SOFM proposed by Kohonen, the neurons are organized in a one-dimensional linear array or a two-dimensional rectangular or hexagonal mesh. Since then, many researchers have made progress in using different cell structures to accommodate various high-dimensional data distributions [8, 9, 10]. As the dimensionality of the training patterns increases and the distribution becomes more unusual, SOFM with one- or two-dimensional neighborhood structure have more difficulty in representing the training set with reasonable number of neurons.

In this paper, with a view to building vector quantization codebooks that have inherent binary tree structure, we propose a modified SOFM learning algorithm, called the Binary-Tree-structured Self-Organizing Feature Maps (BTSOFM). In this algorithm, the definition of "neighborhood" and "distance between neurons" is modified. Figure 1 illustrates a 16-neuron BTSOFM together with the distances between a particular neuron and all other neurons. As can be easily seen the "distance" between two neurons is basically the number of levels from either neuron to their common ancestor in the binary tree. The updating rule of BTSOFM is similar to that of the original SOFM and is listed in Figure 2.

In the early stage of training, BTSOFM partitions the codewords into two groups, consisting of similar codewords with the same mean (dc) values. As the neighborhood is shrunk, the codebook is further divided into more groups with different dc values. In the latter few steps the codewords has more variations and edge-like codewords begin to appear. Along this evolutionary process, the codewords maintain a binary-tree structure. The final codebook has a full eight-level-deep binary tree structure.

When compared with the codebook produced by the LBG algorithm (see Figure 3), one finds that the BTSOFM-generated codebook not only is better-structured but also has more diversity and contains more edge-like codewords. In human visual model, the information carried by edge blocks are more important than that of low-variation blocks. So, the codebook designed by BTSOFM often produced better-quality vector-quantized images than the LBG algorithm does. The reason for the lack of diversity in LBG-designed codebook is that the LBG algorithm computes codewords by averaging the input patterns belonging to the same cluster, which tends to blur the edge features since averaging a kind of low-pass filtering. Unlike the LBG algorithm, in BTSOFM the codewords are updated as individual training pattern is learned. This mechanism together with decreasing learning rate and shrinking neighborhood effectively implements some kind of annealing on the total distortion. Therefore, the learning process can escape local minima in the optimization process and produce better codebook than the gradient-based LBG algorithm.

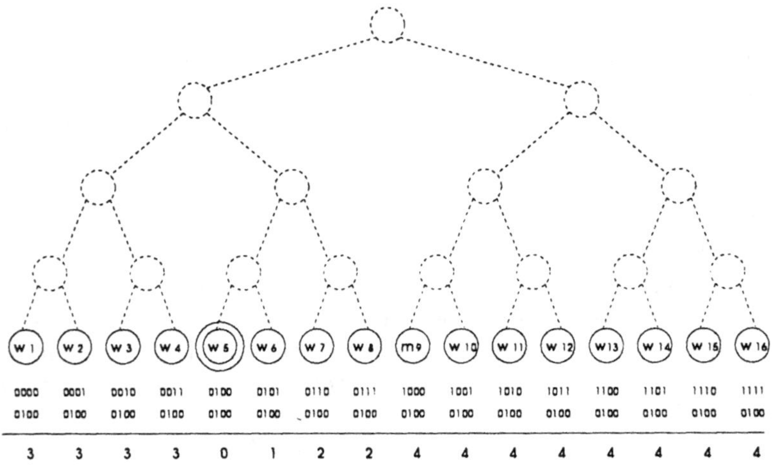

Figure 1: Cell structure of the proposed binary-tree-structure self-organizing feature maps (BTSOFM) model.

BTSOFM Algorithm

(1) Input a training pattern.

(2) Search *all* neurons to find the one with most similar codeword.

(3) Update codewords of all neurons within current neighborhood.

(4) If not in equilibrium, go to step (1). If in equilibrium, then shrink neighborhood, decrease learning rate, and go to step (1). If in equilibrium and the neighborhood has only one neuron, then stop.

Figure 2: The BTSOFM learning algorithm

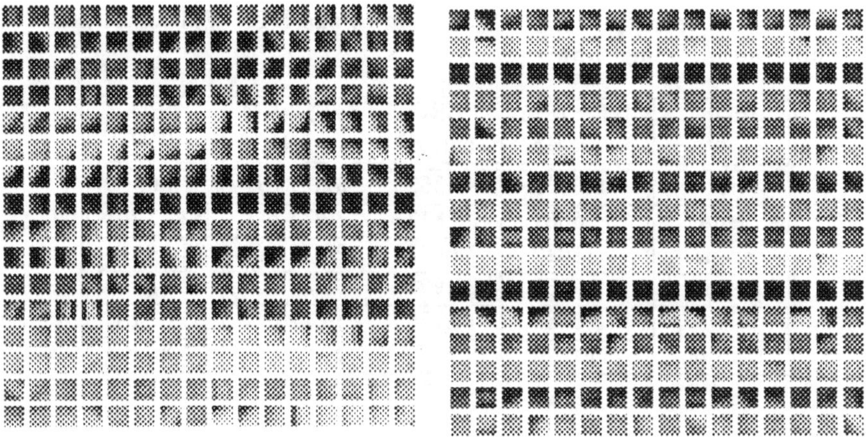

(a) (b)

Figure 3: The final codebook generated by (a) the BTSOFM algorithm and (b) the LBG algorithm

3 Simulation Results

We have used the LBG algorithm and the BTSOFM algorithm to design codebooks for several images. The final mean squared error (MSE) and the signal to noise ratio (PSNR) for these images are listed in Table 1. It is obvious that for most images the BTSOFM algorithm produces a better codebook with lower MSE. Generally speaking, the BTSOFM codebook preserves more edges and reproduces more natural-looking images than the LBG codebook does. For example, in the case of the "lena" image, the BTSOFM-reconstructed image is smoother at the rim of lena's hat and at the shoulder line. To more clearly observe the difference of these two reconstructed images, we show an expanded view of the shoulder area in Figure 4.

Since the BTSOFM algorithm generates codebooks with a binary-tree structure, it should be easier to construct the "decision tree" for finding the nearest codeword. In that case, a search for the nearest codeword requires only $\log_2 256 = 8$ comparisons instead of 256 needed by the full search algorithm. Although the distortion may increase (usually by less than 1 dB) when using the "tree search" mechanism, the saving in encoder complexity is enormous. Moreover, with redundant codewords at leaf nodes of the binary tree, tree-search encoding can yield images that are as good as those coded by the full search method.

Table 1: Comparison of the distortion by the BTSOFM algorithm and the LBG algorithm

image	BTSOFM (MSE)	LBG (MSE)	BTSOFM (PSNR)	LBG (PSNR)
jet	39.0578	42.3623	32.21 dB	31.86 dB
lena	34.5587	36.8132	32.75 dB	32.47 dB
scene	75.1012	79.9324	29.37 dB	29.10 dB
pepper	37.9369	39.9980	32.34 dB	32.11 dB
milkdrop	13.6782	14.1559	36.77 dB	36.62 dB
baboon	217.6185	213.8385	24.75 dB	24.83 dB

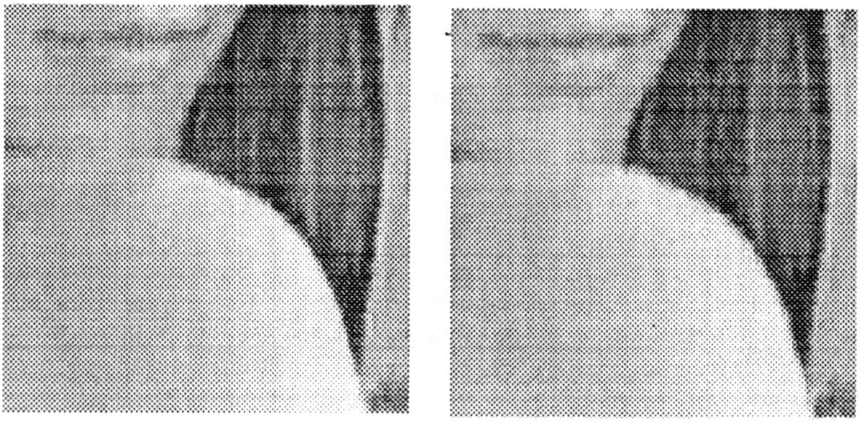

(a) (b)

Figure 4: Expanded view of the reconstructed "lena" images encoded by (a) the BTSOFM codebook and (b) the LBG codebook

4 Conclusions

In this paper, we proposed a modified SOFM algorithm that has a binary-tree-like cell structure. When applied to designing codebooks for vector quantization, the new algorithm generates codebooks with binary tree structure. Experiment results show that the BTSOFM method produces codebooks withmore diversity and more structure than the LBG algorithm.

Acknowledgements

This work is supported in part by the National Science Council, Taiwan ROC under grant No. NSC 82-0404-E-002-224 and by the Telecommunication Labs, Ministry of Communications, Taiwan ROC under grant No. TL-NSC-82-5205.

References

[1] A. Gersho and R.M. Gray, *Vector Quantization and Signal Compression,* Boston, MA: Kluwer Academic Press, 1992.

[2] R.M. Gray, "Vector Quantization," *IEEE ASSP Magazine,* vol. 1, pp. 4–29, 1984.

[3] Y. Linde, A. Buzo, and R.M. Gray, "An Algorithm for Vector Quantizer Design," *IEEE Trans. on Communications,* vol. 28, pp. 84–95, 1980.

[4] D. Cheng and A. Gersho, "A Fast Codebook Search Algorithm for Finding Best Matches in Logarithmic Expected Time," in *Proceedings of ICASSP*, 1986, pp. 265–268.

[5] C.H. Hsieh, "DCT-based Codebook Design for Vector Quantization of Images," *IEEE Trans. on Circuits and Systems for Video Technology,* vol. 2, pp. 401–409, 1992.

[6] E.A. Riskin and R.M. Gray, "A Greedy Tree-Growing Algorithm for the Design of Variable Rate Vector Quantizer," *IEEE Trans. on Signal Processing,* vol. 39, pp. 2500–2507, 1991.

[7] N.M. Nasrabadi and R.A. King, "Image Coding Using Vector Quantization: A Review," *IEEE Trans. on Communications,* vol. 36, pp. 957–971, 1988.

[8] J.A. Kangas and T.K. Kohonen, "Variants of Self-Organizing Maps," *IEEE Trans. on Neural Networks,* vol. 1, pp. 93–99, 1990.

[9] B. Fritzke, "Let It Grow — Self-Organizing Feature Maps with Problem Dependent Cell Structure," in *Proceeding of International Conference on Artificial Neural Networks,* vol. 1, 1991. pp. 403–408.

[10] T. Martinez and K. Schulten, "A Neural-Gas Network Learns Topologies," in *Proceeding of International Conference on Artificial Neural Networks,* vol. 1, 1991, pp. 397–402.

Identifying Fault-Prone Software Modules Using Connectionist Networks

N. Karunanithi

Room 2E-378, Bellcore
445, South Street, Morristown, NJ 07960
karun@faline.bellcore.com

Abstract

Static complexity metrics are often used to measure the functional complexity of software modules. Classifying software modules, based on their static complexity measures, into different fault-prone categories is an important problem in software engineering. This research investigates the applicability of neural network classifiers for identifying fault-prone software modules using a data set from a commercial software system. A preliminary empirical comparison is performed between a minimum distance based Gaussian classifier, a perceptron classifier and a multilayer feed-forward network constructed using a modified Cascade-Correlation algorithm. Our preliminary results suggest that multilayer feed-forward networks can be used as a tool to identify fault-prone software modules early during the development cycle.

1 Problem Statement

Reliability of a software system can be improved by identifying troublesome software modules early during the development cycle. Many measurable program attributes have been used to characterize the intrinsic complexity and the fault proneness of software systems. The intuition behind software complexity metrics is that complex program modules tend to be more error prone than simple modules. By controlling the complexity of software modules during development, one can produce software systems that are easy to maintain and enhance. Static complexity metrics are measured from the passive program texts early during the development cycle and can be used as a valuable feedback for allocating resources in future releases and new projects.

There are two approachs for relating static complexity measures with faults found (or program changes made) during testing. In the *estimative approach* regressions models are used to predict the actual number of faults that will be disclosed during testing [9, 2, 6]. Regression models assume that the metrics that constitute independent variables are i.i.d normals. However, most practical measures often violate these assumptions. The resulting fit of the regression models often tend to produce inconsistent predictions.

In the *classification approach* software modules are categorized into two or more fault-prone classes. A special case of the classification approach is to classify software modules into either low-fault (non-complex) or high-fault (complex) categories. The main rationale behind this approach is that the software managers are often interested in getting some approximate feedback from this type of models rather than accurate predictions of the number of faults that will be disclosed. Existing two-class categorization models are based on linear discriminant principle [8, 6]. Linear discriminant models assume that the metrics are orthogonal and that they follow a normal distribution. To reduce multicollinearity, researchers often use principle component analysis or some other dimensionality reduction techniques. However, the reduced metrics may not explain all the variability if the original metrics have nonlinear relationship.

In this research, the applicability of neural network classifiers for identifying fault proneness of software modules is examined. The motivation behind this research is to evaluate whether classifiers can be developed without usual assumptions about the input metrics. In order to study the usefulness of neural network classifiers, a preliminary comparison is made between a simple minimum distance based Gaussian classifier, a single layer perceptron and a multilayer feed-forward network developed using a modified Cascade Correlation algorithm [3]. The modified algorithm incorporates a cross-validation for constraining the growth of the size of the network. In this investigation, other issues such as selection of proper training samples and representation of metrics are also considered.

2 Data Set Used

The metrics data used in this study were obtained from a research conducted by Lind and Vairavan [5] for a Medical Imaging System software. The complete system consists of approximately 4500 modules amounting to about 400,000 lines of code written in Pascal, FORTRAN, PL/M and assembly level. From this set, a random sample of 390 high level language routines was selected for the analysis. For each module in the sample, program changes were recorded as an indication of software fault. The number of changes in the program modules varied from zero to 98. In addition to changes, 11 software complexity metrics were extracted from each module. These metrics range from total lines of code to Belady's bandwidth metric. (Readers curious about these metrics may refer to Table I of [5].) For the purpose of our classification study, these metrics represent 11 input (both real and integer) variables of the classifier.

A software module is considered as a low fault-prone module (Category I) if there are zero or 1 changes and as a high fault-prone module (Category II) if there are 10 or more changes. The remaining modules are considered as medium fault category. For the purpose of this study we consider only the low and high fault-prone modules. Our extreme categorization and deliberate discarding of program modules is similar to the approach used in other studies [8, 6]. After discarding medium fault-prone modules, there are 203 modules left in the data set. Of 203 modules, 114 modules belong to the low fault-prone category while the remaining 89 modules belong to the high fault-prone category.

3 Training Data Selection

We had two objectives in selecting training data: 1) to evaluate how well a neural network classifier will perform across different sized training sets and 2) to select the training data as much unbiased as possible. The first objective was motivated by the need to evaluate whether a neural network classifier can be used early in the software development cycle. Thus, the classification experiments were conducted using training sets of increasing size $\mathbf{S} = \frac{1}{4}, \frac{1}{3}, \frac{1}{2}, \frac{2}{3}, \frac{3}{4}, \frac{9}{10}$ fraction of 203 samples belonging to Categories I and II. The remaining (**1-S**) fraction of the samples were used for testing the classifiers. In order to avoid bias in the training data, we randomly selected 10 different training samples for each fraction **S**. This resulted in 6 X 10 (=60) different training and test sets.

4 Classifiers Compared

4.1 A Minimum Distance Classifier

In order to compare neural network classifiers and linear discriminant classifiers we implemented a simple minimum distance based two-class Gaussian classifier of the form [7]:

$$|X - C_i| = ((X - C_i)(X - C_i)^t)^{1/2}$$

where C_i, $i = 1, 2$ represent the prototype points for the Categories I and II, X is a 11 dimensional metrics vector, and t is the transpose operator. The prototype points C_1 and C_2 are calculated from the training set based on the normality assumption. In this approach a given arbitrary input vector X is placed in Category I if $|X - C_1| < |X - C_2|$ and in Category II otherwise.

All raw component metrics had distributions that are asymmetric with a positive skew (i.e., long tail to the right) and they had different numerical ranges. Note that asymmetric distributions do not conform to the normality assumption of a typical Gaussian classifier. First, to remove the extreme asymmetry of the original distribution of the individual metric we transformed each metric using a natural logarithmic base. Second, to mask the influence of individual component metric on the distance score, we divided each metric by its standard deviation of the training set. These transformations considerably improved the performance of the Gaussian classifier. To be consistent in our comparison we used the log transformed inputs for other classifiers also.

4.2 A Perceptron Classifier

A perceptron with a hard-limiting threshold can be considered as a realization of a non-parametric linear discriminant classifier. If we use a sigmoidal unit, then the continuous valued output of the perceptron can be interpreted as a likelihood or probability with which inputs are assigned to different classes. The perceptron had two sigmoidal output units (outputs 1 and 2) corresponding to two categories. A given arbitrary vector $\mathbf{X}$ is assigned to Category I if the value of the output unit 1 is greater than the output of the unit 2, and to Category II otherwise. The weights of the network are determined iteratively using least square error minimization procedure. In almost all our experiments, the perceptron learned about 75 to 80 percentages of the training set. This implies that the rest of the training samples are not linearly separable.

4.3 A Multilayer Layer Classifier

To evaluate whether a multilayer network can perform better than the other two classifiers, we repeated the same set of experiments using feed-forward networks constructed using the Cascade-Correlation algorithm [3]. The Cascade-Correlation algorithm is a constructive training algorithm which constructs a suitable network architecture by adding one hidden (layer) unit at a time. Our initial results suggested that the multilayer layer networks constructed by the Cascade-Correlation algorithm are not capable of producing a better classification accuracy than the other two classifiers. An analysis of the network suggested that the resulting networks had too many free variables (i.e., due to too many hidden units). A further analysis of the rate of decrease in the residual error versus the number of hidden units added to the networks revealed that the Cascade-Correlation algorithm is capable of adding more hidden units to learn individual training patterns at the later stages of the training phase than in the earlier stages. This happens if the training set contains patterns that are interspersed across different decision regions or what might be called "border patterns" [1]. In an effort to constrain the growth of the size of the network, we modified the Cascade-Correlation algorithm to incorporate a cross-validation check during the output layer training phase. For each training set of size **S**, one third was used for cross-validation and the remaining two third was used to train the network. The network construction was stopped as soon as the residual error of the cross-validation set stopped decreasing from the residual error at the end of the previous output layer training phase. The resulting network learned about 95% of the training patterns. However, the cross-validated construction considerably improved the classification performance of the networks on the test set.

5 Results

In this section we present some preliminary results comparing different classifiers in terms of classification accuracy. Since a neural network's performance can be affected by the weight vector used to initialize the network, we repeated the training experiment 25 times with different initial weight vectors for each training set. This resulted in a total of 250 training trials for each value of **S**. The results reported here for the neural network classifiers represent a summary statistics for 250 experiments. The performance of the classifiers are reported in terms of classification errors. There are two type of classification errors that a classifier can make: a Type I error occurs when the classifier identifies a low fault-prone (Category I) module as a high fault-prone (Category II) module; a Type II error is produced when a high fault-prone module is identified as a low fault-prone module. From a software manager's point of view, these classification errors will have different implications. Type I misclassification will result in waste of test resources (because modules that are less fault-prone may be tested longer than what is normally required). On the other hand, Type II misclassification will result in releasing products that are of inferior quality. From reliability point of view, a Type II error is a serious error than a Type I error.

No. of Patterns			Error Statistics					
S	Training	Test	Gaussian		Perceptron		Multilayer Nets	
%	Set	Set	Mean	Std	Mean	Std	Mean	Std
Type I Error Statistics								
25	50	86	13.16	4.7	16.17	5.5	20.19	5.4
33	66	77	11.44	4.0	11.74	3.9	18.24	5.5
50	101	57	12.45	3.2	11.58	3.2	17.41	5.6
67	136	37	9.46	4.1	10.14	3.9	14.32	5.8
75	152	28	8.57	5.4	9.15	5.8	13.27	7.0
90	182	12	14.17	7.9	4.03	4.3	9.77	9.4
Type II Error Statistics								
25	50	67	15.61	4.2	15.98	7.8	12.11	4.7
33	66	60	15.46	4.6	15.78	6.6	12.40	4.1
50	101	45	16.01	5.1	16.97	6.8	15.04	5.2
67	136	30	16.00	5.4	16.11	7.6	14.08	5.5
75	152	23	17.39	5.8	18.39	6.3	13.84	5.4
90	182	9	21.11	6.3	19.11	5.6	15.47	5.1

Table 1: A Summary of Type I and Type II Error Statistics.

Table 1 illustrates preliminary results for different classifiers. The first two columns represent the size of the training set in terms of **S** as a percentage of all patterns and the number of patterns respectively. The third column represents the number of test patterns in Categories I (1st half) and II (2nd half). The remaining six columns represent the error statistics for the three classifiers in terms of percentage mean errors and standard deviations. The percentages errors were obtained by dividing the number of misclassifications by the total number of test patterns in that category. The Type I error statistics in the first half of the table suggest that the Gaussian and the Perceptron classifiers may be better than multilayer networks at early stages of the software development cycle. However, the difference in performance of the Gaussian classifier is not consistent across all values of **S**. The neural network classifiers seem to improve their performance with an increase in the size of the training set. Among neural networks, the perceptron classifier seems to perform classification than a multilayer net. However, the Type II error statistics in the second half of the table suggest that a multilayer network classifier may provide a better classification of Category II modules than the other two classifiers. *This is an important results from the reliability perspective.*

6 Conclusion

We demonstrated the applicability of neural network classifiers for identifying fault-prone software modules. We compared the classification efficacy of three different pattern classifiers using a data set from a commercial software system. Our preliminary empirical results are encouraging in that there is a role for multilayer feed-forward networks either during the software development cycle of a subsequent release or for a similar product.

References

[1] Ahmed, S. and Tesauro, G., "Scaling and Generalization in Neural Networks: A Case Study", *Advances in Neural Information Processing Systems 1*, 1989, pp 160-168, D. Touretzky, ed., Morgan Kaufmann.

[2] Crawford, S. G., McIntosh, A. A. and Pregibon, D., "An Analysis of Static Metrics and Faults in C Software", *The Journal of Systems and Software*, Vol. 5, pp. 37-48, 1985.

[3] Fahlman, S. E. and Lebiere, C., "The Cascaded-Correlation Learning Architecture," *Advances in Neural Information Processing Systems 2*, 1990, pp 524-532, D. Touretzky, ed., Morgan Kaufmann.

[4] Karunanithi, N., Whitley, D. and Malaiya, Y. K., "Prediction of Software Reliability Using Connectionist Models", *IEEE Trans. on Software Eng.*, Vol. 18, No. 7, pp. 563-574, July 1992.

[5] Lind, R. K. and Vairavan, K., "An Experimental Investigation of Software Metrics and Their Relationship to Software Development Effort", *IEEE Trans. on Software Eng.*, Vol. 15, No. 5, pp. 649-653, May 1989.

[6] Munson, J. C. and Khoshgoftaar, T. M., "The Detection of Fault-Prone Programs", *IEEE Trans. on Software Eng.*, Vol. 18, No. 5, pp. 423-433, May 1992.

[7] Nilsson, J. Nils, **The Mathematical Foundations of Learning Machines**, Morgan Kaufmann, 1990, Chapters 2 and 3.

[8] Rodriguez, V. and Tsai, W. T., "A Tool for Discriminant Analysis and Classification of Software Metrics", *Information and Software Technology*, Vol. 29, No. 3, pp. 137-149, Apr. 1987.

[9] Shen, V. Y., Yu, T., Thebaut, S. M. and Paulsen, L. R., "Identifying Error-Prone Software: An Empirical Study", *IEEE Trans. on Software Eng.*, Vol. SE-11, No. 4, pp. 317-323, Apr. 1985.

Clone Detection in Telecommunications Software Systems: A Neural Net Approach

S. Carter, R.J. Frank*, D.S.W. Tansley

BNR Europe Limited
London Road, Harlow, Essex, UK, CM17 9NA
sc@bnr.co.uk, dswt@bnr.co.uk
* *University of Hertfordshire*
College Lane, Hatfield, Herts, UK, AL10 9AB
comqrjf@herts.ac.uk

Abstract

We report on the development of a tool which uses neural nets to help in the detection of 'clone' software in large telecommunication software systems. Cloning is a primitive form of software re-use, whereby existing blocks of source code are copied and adapted, for use elsewhere in the system or to implement new functionality. A number of representation schemes for presenting the source code to Self-Organising Maps (SOMs) have been tried. The more favourable of these have been applied in a hybrid, modular tool architecture. The current result is a demonstration system in trial.

1 Background

1.1 Work context

The authors were given the opportunity to start to evaluate the costs and benefits of the much-hyped neural network technology in late 1992, following years of low-level background monitoring of the field by the Software and Systems Engineering group within the Advanced Technology Centre at BNR Europe Limited, Harlow, UK. Our aim was to design a programme of work to be carried out in 1993 to demonstrate the capabilities of the technology in a useful way to the company. It was decided that this should be done by developing a software source code clone detector.

1.2 Clones and cloning

The clone 'problem[1]' arises from a simple, frequently-used technique to increase or amend the functionality of telecommunication systems, notably those large, older systems from a time when software development techniques were relatively primitive compared to those we have today. It involves copying existing sections of source code and changing the copy as necessary in order to implement the new functionality. Whilst this should result in copying known-to-be-working code, it also copies as-yet-unknown bugs, and may often include large amounts of redundant statements. Software cloning is a major contributor to the steadily increasing size, complexity, and increased error rates of such systems.

The choice for telecommunications companies is stark – either build new systems from scratch using modern technology, or re-engineer the existing systems. If re-engineering is the choice, a major part of the problem will be solved if the clone software present in the system can be managed – either eradicated or tracked. Similarly, new clones should either be prevented or logged to aid future bug-fixing. To do this practically, an automatic tool is required – a clone detector. Since clone detection inevitably involves pattern matching, a neural net approach would seem to offer some promise.

There are a number of different types of clone to be found in telecommunications software. By different types, we mean degrees of how far away from some original piece of software a cloned piece of code is. Thus, one can describe clones using the following typology:

- Type I – An exactly identical source code clone, i.e. no changes at all.
- Type II – An exactly identical source code clone, but with indentation, comments, or identifier (name) changes.
- Type III – A clone with very similar source code, but with small changes made to the code to tailor it to some new function.
- Type IV – A functionally identical clone, possibly with the originator unaware that there is a function already available that accomplishes essentially the same function.

[1]There are actually both positive and negative aspects of software cloning.

Whilst a neural net approach could be used to address any of these types, currently we are addressing the problem of types II and III, which probably form by far the majority of clones present in existing systems. Future work may address type IV. (Type I is trivial to detect.)

Finally, in terms of level of scale, at the current stage of our work, we are looking for clones at the 'procedure' or 'PROC[2]' level. (This is analogous to the level of 'functions' in the C programming language.)

2 A Neural Network Solution

The problem of detecting clones is to do with taking a large data set and identifying groups of closely-related sections of that data, i.e. a classification problem. The data set is some range of the total set of source code procedures for a system that we are concerned with. By 'closely related', we mean 'closely related in terms of being candidate clones'.

A solution emerged in the form of the well-known 'Self-Organising Map' (SOM) [1] [2] . This performs a classification using an unsupervised training phase. It had the additional advantage of preserving any topological relationships between the input vectors. This would be useful in our problem domain as it might be used to measure some degree of 'cloniness' between any specific classes of clones identified by the network.

3 The Input Vector

The mapping of the raw source code into a form suitable for inputting to the network is a major area for our investigation. Preliminary discussions generated the following possibilities for abstracting the raw source code into its specific features: call graphs, parse trees, tokens (e.g. names, identifiers, etc.), visual patterns (e.g. indentation, characters represented as single bits, etc.), data structures, and comments. The input to the net would be some coded form of these. Combinations of the above are also possibilities.

The precise form of the input would be dependent upon the specific type(s) of network used. Our early experiments concentrated on two features of the source code in particular: physical layout patterns, keyword and operator frequency distribution, (and also combinations of these two). Although first attempts, these have actually proved successful enough to not warrant the trying of other features at this stage.

Figure 1 illustrates how a typical input vector is constructed.

[2]A statement in BNR's proprietary programming language – but note that our work is *completely* language-independent.

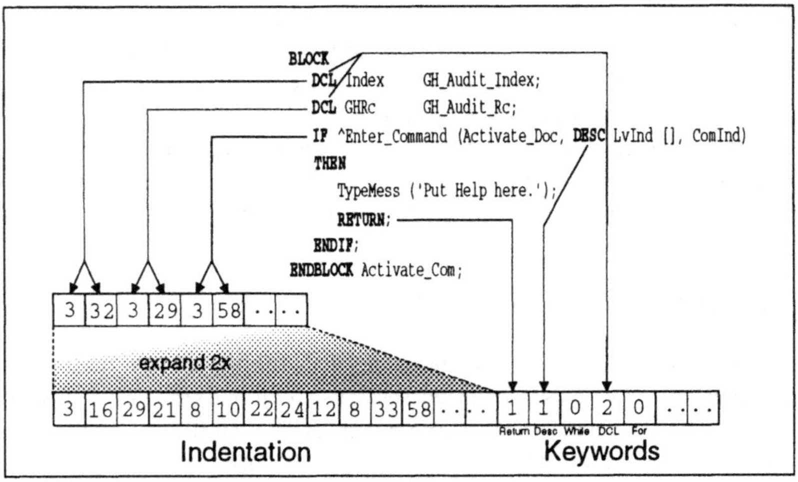

Figure 1: Schematic of input vector construction

The first type of source code feature we used was the 'physical' way the source code was layed out, i.e. how it was indented on the page, how many lines were present, how long the lines were, etc. There are various possible combinations of these elements to describe the feature of the way the code is laid out, depending on the level of detail required and the relative importance of each element. For example, we tried representing only the changes in indentation as a binary pattern vector and we tried representing the changes and the length of similarly indented blocks.

A significant problem with this approach was that the vector varied in length depending upon how many changes in indentation there were and/or how long the procedure was. We used padding and truncation to address these problems initially, in order to get a fixed input vector for the SOM, but later moved to a scaling algorithm to pre-process the data to always produce a fixed-length vector. The scaling algorithm worked by using the values in the original vector to define points on a graph with straight lines joining those points. The new vector was produced by sampling along this graph at the required interval to find the values for the new vector. This technique has the advantage that operations such as adding or removing a line of source code would result in similar vectors.

The left-hand side of figure 1 shows how a real vector (part) is constructed for indentation (and line length). For each line of code, the number of characters of indentation is one value of a data pair (e.g. 3); the other value is the length of the line (e.g. 32). In the example, a x2 expansion is required to scale the raw data vector to the fixed-length input vector.

The second type of source code feature we used was the frequency distribution of the programming language's keywords used in the procedures. Keywords were such things as 'RETURN', 'DESC', 'WHILE', 'IF', 'THEN', etc. There were 77 different keywords, although some of these could be removed due to rarity of use or redundancy (due to them always appearing in pairs, for example), leaving 62 meaningful ones. The keyword input vector (part) thus simply consisted of one fixed-length array per procedure, with each element of the array containing a number representing the count of occurrences of a particular keyword in the current procedure – see right-hand side of figure 1. Additionally, we added counts of the 19 different operators in the language, giving a total input vector size of 81 in this case.

4 The Output Vector

The output of a SOM is inherently coupled to its Kohonen layer – in fact, it is the pattern of activity in that layer itself. When a vector is applied to the trained network, one neuron becomes active in the Kohonen layer. Each neuron in the Kohonen layer represents a class, so the active neuron indicates the class that the input vector falls within. This raw output is not suitable in a clone detector tool. The sort of output required in such a tool is a list of 'candidate clones', perhaps with degrees of likelihood or 'cloniness' against each, compared to themselves or a 'reference procedure'. This output information can easily be obtained from the SOM with a little post-processing, however, and its topology-preserving characteristic helps to make a 'cloniness' measure more easy to calculate.

5 Experimental Results

These results were produced using input vectors from 1775 procedures extracted from a skim file of approximately 5Mbytes of arbitrarily-selected source code. A SOM was trained using these vectors. At the end of training the output from the SOM was further processed to populate a clone database. Different experiments were done using combinations of various sized SOMs and different input vectors (see Section 3).

The clone detection rate was determined by randomly choosing 100 test procedures from this data set and using the clone database to access the nearest procedure to it, subject to a cloniness threshold. The two procedures, test and candidate clone, were viewed side by side by a software engineer familiar with the type of source code who then decided whether the neighbouring procedure was or was not a clone of the test procedure. A numerical value was then arrived at for each test by examining the result for each procedure in the test and adding the cloniness value if the procedure was a clone and subtracting the cloniness value if it was not a clone. The final result was then scaled to be in the range 0 to 100. Note that this is therefore *not* a percentage measure of accuracy!

Table 1 summarises the key experiments performed with the different input vector schemes described in Section 3.

Table 1: Input Vector Type vs Kohonen Layer Size Results

Input Vector Scheme	**Kohonen Layer Size**	**Results**
Keyword/Operators	10x10	35.2
Keyword/Operators	15x15	36.0
Keyword/Operators	20x20	37.0
Indentation	10x10	41.3
Indentation	15x15	51.4
Indentation	20x20	55.0
Keywords/Operators/Indentation	20x20	68.7

The results show that as the number of units increase in the Kohonen layer, the categorisation becomes more accurate.

The results also show that although the keyword and indentation techniques both produce reasonable results, the combination of the two produces a better result, as more information about the procedure is provided to the net and there is a smaller likelihood that two procedures that were not clones of each other would have similar input vectors.

6 The Neural Network Clone Detection Tool

Figure 2 shows a schematic view of the proposed tool architecture and mode of operation. This is described below.

The proposed tool will be required to operate with a much larger environment (100k+ procedures) compared to the development and test environment. We plan to use a two level hierarchical neural network based on the SOM paradigm. The first level SOM will be trained on all the procedures (in some suitable subset of the entire system) to classify them into a relatively small number, say n, classes.

A set of n separate second level SOMs will then be trained on the procedures in each class to discriminate these broad classes into smaller sub-

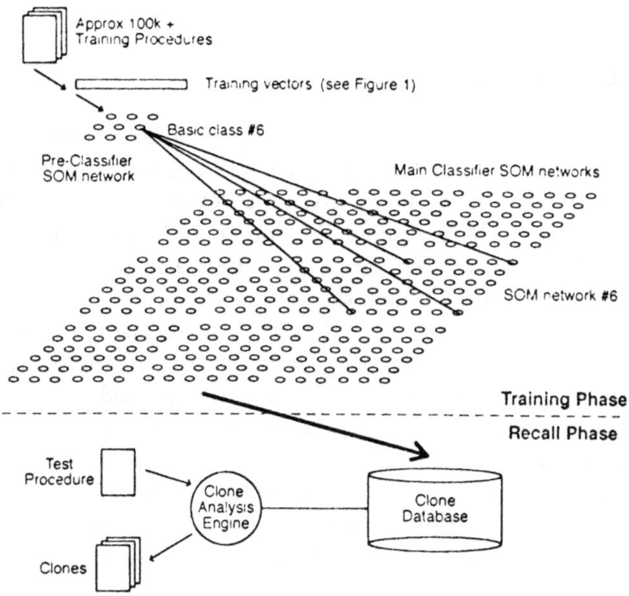

Figure 2: Schematic of clone detection tool

classes. The output of this neural network hierarchy would then be processed into a cloned procedure database. This would be accessed by the clone analysis component of the tool in a recall phase, once the training had populated the database.

The clone detection tool would find the clones of a given procedure by accessing the class and subclass of a procedure from the neural network output database and calculating the closeness of the given procedure to the other procedures in the sub-class and neighbouring sub-classes.

At this point the closeness measure used may be based upon the textual differences of the source code, procedure names, etc. The user can inspect any of the candidate clone procedures against the test procedure in text windows placed side by side on a workstation.

This mode of operation of the tool will be useful in debugging where the cloned procedures of a procedure with a known error can also be discovered, allowing an increase in efficiency of this activity.

Another possible mode of operation of the tool would be in providing a statistical analysis of large parts of the system. This would constitute a more time-consuming operation but would provide system development managers with an overall view of the state of the system from the point of view of code duplication, development effort, etc.

The tool could also be used as an aid in producing a reusable software library for use in later generations of the software.

7 Conclusions

A demonstration tool has been produced that will take the raw output from a trained SOM, together with the training vectors, in order to produce an ordered list of likely clones of any given procedure. It will then present windows on a screen which display the source code containing the likely clones in order for the user to inspect them manually.

The work as it currently stands shows that the best performing network is one that has the order of a 20x20 Kohonen layer, using an input vector that encodes a combination of source code features including indentation pattern, keyword and operator frequency, and the procedure length.

A clone detector product will use a hierarchy of SOMs which will enable us to manage a realistic amount of source code. This will result in an extremely powerful and uniquely flexible tool. The clone detector will be just the start of a possible portfolio of reverse engineering and design recovery tools that could use components based on neural networks.

Acknowledgments

We would like to thank Neil Davey from the Department of Computer Science at the University of Hertfordshire, UK, for his contributions to the work presented in the paper in his role as consultant.

We also acknowledge the financial assistance of the Department of Trade and Industry of the UK government, who have partly funded the placement of Ray Frank on a 'Senior Academics in Industry Scheme' (SAIS) at BNR Europe Limited, through the Teaching Company Directorate (TCD) organisation, under grant no. SAIS-003.

References

[1] T. Kohonen, "Self Organised formation of topologically correct feature maps," *Biol. Cyber.*, vol. 43, pp. 59–69, 1982.

[2] T. Kohonen, "A simple paradigm for the self-organized formation of structured feature maps," in Amari,S., and Arbib, M.[Eds], *Competition and Cooperation in Neural Nets,* Lecture Notes in Biomathemetics vol. 45, Springer Verlag, 1982.

Learning Index Rules and Adaptation Functions for a Communications Network Fault Resolution System

Lundy Lewis and Steve Sycamore

Cabletron Systems, Inc.
9 Executive Park Drive
Merrimack, NH 03054
lewis@ctron.com
sycamore@ctron.com

Abstract

Case-based reasoning (CBR) operates by remembering, adapting, executing, and storing episodes of problem-solving in a case library. Two requirements for CBR systems are (i) a set of retrieval rules and (ii) a set of adaptation functions. Current practice is to handcraft index rules and adaptation functions manually. This approach is time-consuming and lacks ability to grow beyond its initial implementation. In the paper we describe preliminary work on the development of ways to learn the rules and functions dynamically by "watching" a case library. The application domain is a communications network fault resolution system.

1 Introduction

A recent approach to the representation of communications network fault resolution knowledge is case-based reasoning (CBR) [7]. The basic idea of CBR is to recall, adapt, and execute episodes of former problem-solving in an attempt to deal with a current problem. Former episodes of problem-solving are represented as cases in a case library. When confronted with a new problem, a CBR system retrieves a similar case and tries to adapt the case to solve the outstanding problem. The experience with the proposed solution is embedded in the library for future reference. Unlike rule-base expert system, the goals of CBR systems are to increase knowledge with experience, and to adapt existing knowledge to suggest solutions to unfamiliar problems.

Figure 1 shows a standard network "trouble ticket". Importantly, the trouble ticket can be interpreted as a case, and a collection of such tickets is essentially a case library. Thus, trouble ticket systems are good vehicles for enhancement with CBR techniques in order to build a network fault resolution system.

Entry-ID	00000000116	Submitter	SPECTRUM
Alarm ID	57	Create-date	07/24/92 08:51:08
Alarm Date/Time	07/24/92 08:51:07	Notify-method	O None O Notifier O E-mail
Condition	O Red O Orange O Yellow	Assigned-Priority	O Low O Medium O High
Device Name	Ethernet-Randy	Assigned-to	
Device Type	Ethernet	Last-modified-by	SPECTRUM
IP Address		Modified-date	07/24/92 08:51:08
Trouble	file_transfer_throughput=slow		
Additional Data	number_of_active_ports=20, collision_rate=15, deferment_rate=20, users=31		
History of Trouble			
Probable Cause			
Resolution			
Resolution Status	O Good O No Good O In Progress		
Ticket Status	O New O Assigned O Rejected O Closed		

Figure 1: A Typical Network Trouble Ticket

Two requirements for CBR systems are (i) a set of index rules used to guide the selection of useful prior cases and (ii) a set of functions used to adapt old cases to fit new problems. Figure 2 shows the overall CBR architecture. Current practice is to handcraft index rules and adaptation functions manually. Although safe, this approach is time-consuming and difficult. In the paper we describe preliminary work on the development of ways to learn the rules and functions dynamically by "watching" a case library.

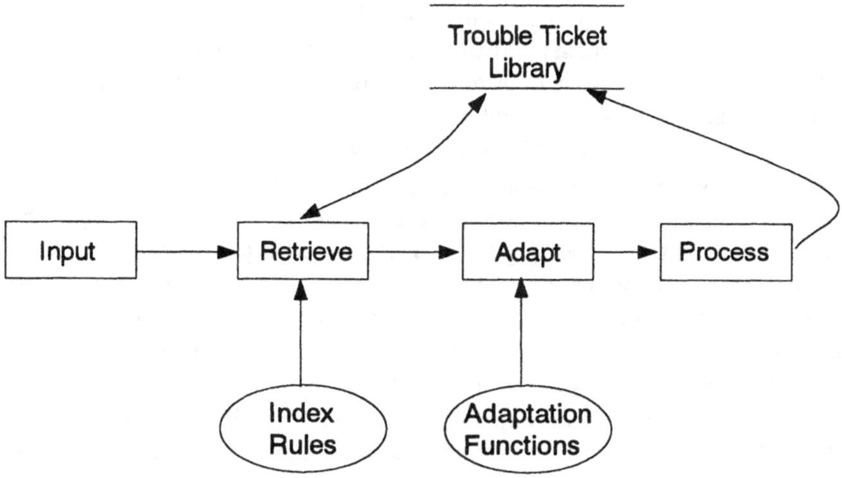

Figure 2: The General CBR Architecture

2 Learning Index Rules

The first challenge for the CBR approach is to develop a similarity metric by which the system can retrieve useful cases from a database. We would not want the similarity metric to be simply "the largest number of matches" on all fields shared between an old case and a new problem. Looking at Figure 1, it is clear that some information, although useful for other purposes, is irrelevant for problem resolution. Thus, we need some way to focus on a subset of the fields when doing the matching. Index rules allow us to do this. An index rule tells the system which fields to consider given certain kinds of problem. Thus, the metric "the largest number of matches on the fields dictated by an index rule" works. An example of an index rule is:

> *If the trouble is "file_transfer_throughput is slow" then consider only bandwidth, number_of_active_ports, packet collision rate, and packet deferment rate*

The system would retrieve the cases with the largest number of matches with respect to the consequent of the rule. Note that an index rule is not the same as a rule in a rule-based expert system. An index rule simply tells the system what cases to look at, but not what to do with them.

The task now is to acquire a good set of index rules. A safe, practical method is to handcraft and test them manually. However, a better approach would be to build a system that can discover its own index rules by examining an existing case library. An additional requirement is to have the system modify its set of index rules over time as new resolved cases are added to the library.

We have examined two machine learning algorithms for this task, an iterative dichotomizing third (ID3) algorithm and the back-propagation (BP) algorithm. Good explanations of ID3 are [1], [2], and [10]. Good explanations of BP are [6], [8], and [12]. An important difference between ID3 and BP is the form of representation of learned knowledge. ID3 represents knowledge as a decision tree or, equivalently, as a set of rules. BP represents knowledge as patterns of activity encoded as connection weights over a set of neurons.

Comparative performance experiments with ID3 and BP are reported in [3], [9], and [13]. The results of the experiments are in general agreement. In non-noisy domains ID3 is preferable to BP because (i) the slight increase in recall accuracy afforded by BP is not worth the extra expense of learning and recall time, (ii) the set-up time for BP is considerably more expensive, (iii) ID3 is simpler and easier to implement, and (iv) ID3 can provide explanations which are difficult to obtain using BP. In noisy domains with continuous input, BP is superior to ID3.

Given these results and the nature of the CBR retrieval task, we are starting our work with an ID3-based algorithm. A related task is reported in [4]. In this work, an ID3-based algorithm is used to learn rules that are implicit in network management databases. The rules are loaded into an expert system shell. However, the work in [4] differs from our approach in that the reasoning paradigm is rule-based, whereas the reasoning paradigm in our approach is case-based.

3 Learning Adaptation Functions

A second challenge for the CBR approach is to develop adaptation techniques by which the system can apply an old solution to a new problem, although the new problem might not be exactly like the old problem. Several such techniques are available [11]. The technique discussed here, parameterized adaptation, is the most useful albeit the most challenging.

Consider an outstanding trouble with file_transfer_throughput, and imagine that only one source case is retrieved from the case library, as shown below. In the example, the resolution is adjust_number_of_active_ports=A; where A is a solution variable that holds a number by which to increase or decrease the total number of active ports.

trouble: file_transfer_throughput=F
resolution: A=f(F), adjust_number_of_active_ports=A
resolution status: good

Parameterized adaptation is a method by which the solution variable of an outstanding case is adjusted relative to its problem variables, based on the relation between the solution and problem variables in a source case. Everything else being equal, an outstanding trouble file_transfer_throughput=F* should propose the resolution adjust_number_of_active_ports=A*, where F* and A* stand in the same relation as F and A in the source ticket. The proposed resolution to the target case, therefore, would look like the following:

*trouble: file_transfer_throughput=F**
resolution: A=f(F*), adjust_number_of_active_ports=A**
resolution status: good

For example, if F and A in the source were 30 KBytes/s and 10 respectively, and if F* in the target case were 33 KBytes/s, then A* would be given the value 11. *f* would be defined as X/3=*f*(X).

The task is to acquire a good set of functions like *f*, where the input to the functions is likely to be multi-dimensional. Current practice is to handcraft and test them manually. A better solution, however, is to have the system discover its own adaptation functions by examining a library of prior cases and to modify its functions over time. This task can be considered as function approximation, and thus the problem of learning the function lends itself to connectionist methods which are good at function approximation, e.g. BP and counter-propagation (CP) [5]. ID3 would not be good at this task because the data is continuous and likely to be noisy. Connectionist algorithms are better equipped to handle the problem, and we begin our work here.

4 Conclusion

CBR techniques are applicable to standard network troube ticket systems in order to effect a network fault resolution system. We have presented preliminary work on two challenges for the CBR paradigm that require machine learning: (i) learning index rules to guide the recollection of old cases for new problems and (ii) learning adaptation functions for adapting old cases to fit new problems. For the former task we will begin our experiments with ID3-based algorithms. For the latter task we will start comparing connectionist algorithms, including BP and CP.

References

[1] B. Arbab and D. Michie , "Generating Rules from Examples," in *Proceedings of the Eleventh International Joint Conference on Artificial Intelligence,* California: Morgan Kaufman Publishers, 1985, pp. 631-633.

[2] J. Carbonel and P. Langley, "Machine Learning," in *The Encyclopedia of Artificial Intelligence,* Wiley and Sons, 1987, pp. 470-472.

[3] D. Fisher and K. McKusic, "An Empirical Comparison of ID3 and Back-Propagation," in *Proceedings of the Eleventh International Joint Conference on Artificial Intelligence,* California: Morgan Kaufman Publishers, 1989, pp. 788-793.

[4] R. Goodman and H. Latin, "Automated Knowledge Acquisition from Network Management Databases," in *Integrated Network Management,* II (editors: I. Krishnan and W. Zimmer), North Holland: Elsevier Science Publishers, 1991, pp.541-549..

[5] R. Hecht-Nielsen, "Applications of Counterpropagation Networks.", *Neural Networks,* Vol. 1, 1988, pp. 131-139.

[6] G. Hinton, "Connectionist Learning Procedures," Technical Report CMU-CS-87-115, Carnegie-Mellon University, Pittsburg, PA., December 1987.

[7] L. Lewis, "A Case-Based Reasoning Approach to the Resolution of Faults in Communications Networks," in *Proceedings of IEEE INFOCOM '93,* IEEE Computer Society Press, pp. 1422-1429.

[8] R. Lippman, "An Introduction to Computing with Neural Nets," *IEEE Acoustics, Speech, and Signal Processing (ASSP) Magazine,* Vol. 4, No. 2, April 1987, pp. 4-22.

[9] R. Mooney, J. Shavlik, G. Towell, and A. Gove, "An Experimental Comparison of Symbolic and Connectionist Learning Algorithms," in *Proceedings of the Eleventh International Joint Conference on Artificial Intelligence,* California: Morgan Kaufman Publishers, 1989, pp. 775-780.

[10] J. Quinlan, "Learning Efficient Classification Procedures and Their Application to Chess End Games," In *Machine Learning: An Artificial Intelligence Approach* (editors: R. Michalski, J. Carbonell, and T. Mitchell), California: Morgan Kaufman Publishers, 1986, pp. 463-482.

[11] C. Riesbeck and R. Schank, *Inside Case-Based Reasoning,* N.J.: Lawrence Erlbaum Associates, 1989.

[12] D. Rumelhart, G. Hinton, and R. Williams, "Learning Internal Representations by Error Propagation," in *Parallel Distributed Processing* (vol. 1), Cambridge: MIT Press, 1988, pp. 318-362.

[13] S. Weiss and I. Kapouleas , "An Empirical Comparison of Pattern Recognition, Neural Nets, and Machine Learning Classification Methods," in *Proceedings of the Eleventh International Joint Conference on Artificial Intelligence,* California: Morgan Kaufman Publishers,1989, pp. 781-787.

Using Distributed Neural Networks to Identify Faults in Switching Systems

Tadashi Sone

NTT Communication Switching Labs,
9-11,Midori-Cho 3-Chome,Musashino-Shi,
Tokyo 180 Japan
sone@luna.ntt.jp

Abstract

A fault identification system that employs a distributed neural network architecture for switching system diagnosis is described. This distributed neural network structure consists of several micro-viewing networks and one macro-viewing network, that take into account the diagnosis control flow and switching system architecture. The module for integrating the output of these distributed networks employs an expert system. This diagnosis system provides a higher diagnosis rate than a single neural network.

1 Introduction

Electronic switching systems(ESSs) have become very reliable and the failure rate has decreased to where it is difficult to acquire experimental and heuristic knowledge on maintenance. Difficult failures like intermittent failures remain, when an infrequent but difficult failure occurs, inexperienced maintenance engineers cannot efficiently troubleshoot the fault [1][2][3][4]. Therefore, we first developed an efficient dummy fault generator to generate a large amount of training data as experimental and heuristic knowledge on actual switching systems. Then, we proposed a fault identification system that applies a distributed neural network architecture. In practical use, the fault identification system is combined with the dummy fault generator to obtain a large volume of high quality training data (Figure 1).

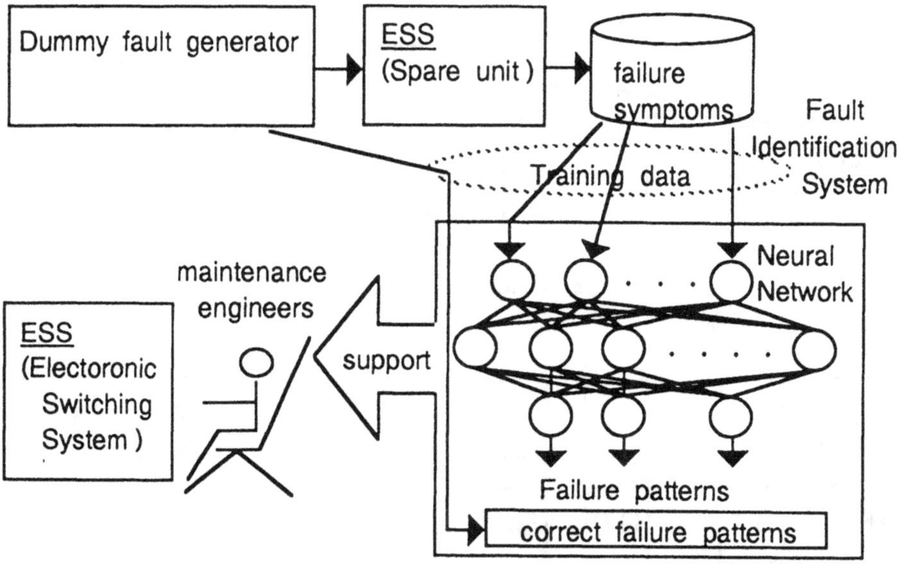

Figure 1: Basic configuration of fault identification system

2 Characteristics of ESS failure

When an ESS fault occurs, many components are suspect. Thus a troubleshooting program needs to replace a suspect component with a properly operating one so as to automatically locate the fault. For example, consider the simple equipment shown in Figure 2. Suppose that components A0 and A1 have a common structure and function, and that B0 and B1 also have a common structure and function. Initially, A0-B0 are the active components and A1-B1 are stand-by units.

When a fault occurs between, for example A1 and B0, as show in Figure 3, a troubleshooting program replaces the active components one-by-one in order, like A0-B0, A0-B1, A1-B1, A1-B0, until the fault is located.

However, the program cannot always detect the faulty component. Therefore, maintenance engineers examine the failure symptoms from the ESS, as shown in Figure 4. These time-series symptoms include information on four states (T1 to T4 in Figure 4). In view of this characteristic of difficult

ESS failures, we proposed the application of the distributed neural network architecture.

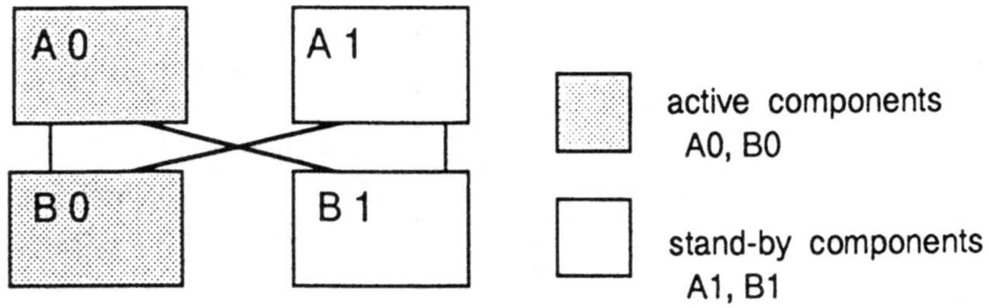

Figure 2: Example equipment

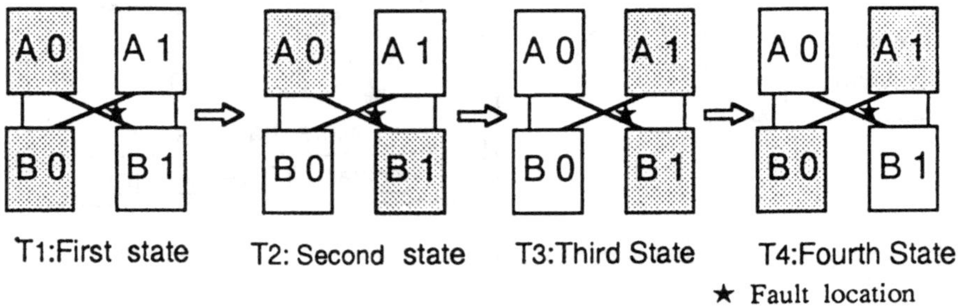

Figure 3: Transitions in machine configuration

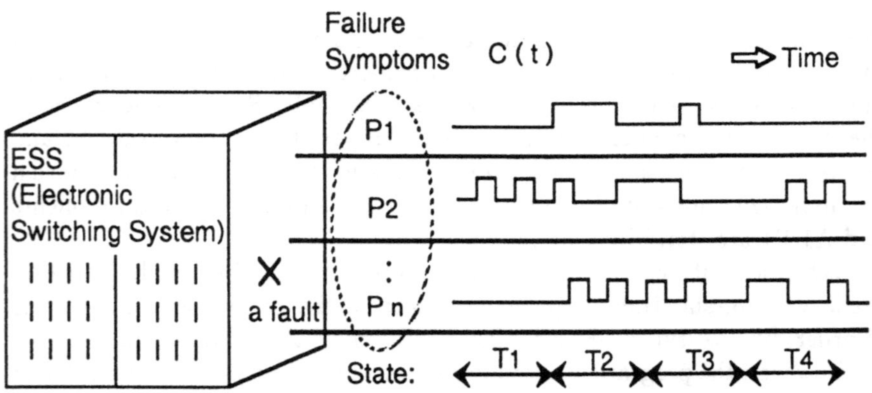

Figure 4 : Time series of sensor signals

3 Distributed Neural Networks Structure

The input to the fault identification system using neural networks is the time-series symptoms, including some state information, as shown in Figure 4. The output is failure patterns that support maintenance engineers in detecting the actual faulty component. Data on 10 failure patterns were experimentally collected using 32 sensor signals from the ESS. The value of each sensor signal is either 0 or 1, and it is updated each second. The interrelations among the signal changes include important information for analyzing failure patterns. The failure patterns were studied by human experts and shown to occur correctly by a dummy fault generator. They were selected as shown in Figure 5 by focusing on given to the importance of the ESS function and the occurrence frequency.

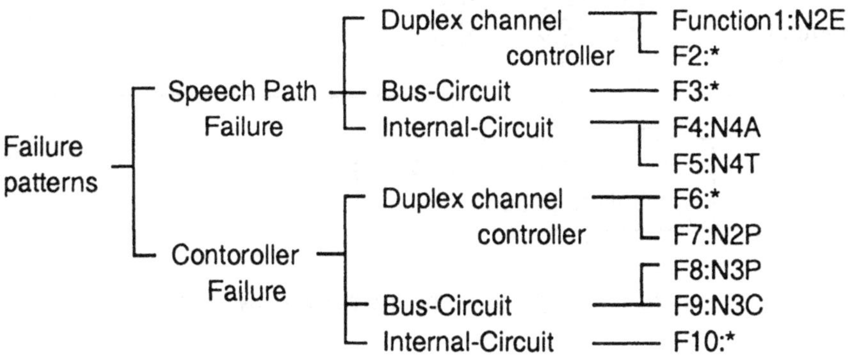

Figure 5 : Failure patterns

As a basic neural network architecture, we use the traditional backpropagation algorithm on a single, three-layer network (input layer, hidden layer, output layer). Time series date is divided among the 30 nodes of the input layer. So, 32 signals require a total of 960 nodes in the input layer (Figure 6).

Initially, we used a single large neural network with the time series data for each signal divided among 30 nodes, for a total of 960 input nodes. The highest recognition rate achieved with this condition was 58%. To improve learning and generalization, we replaced the single network with a distributed network comprising several small networks. The distributed network consists of several micro-viewing networks and one macro-viewing network. Each micro-viewing network analyzes its respective state of the time series in detail, and the macro-viewing network analyzes all four of the states of time series roughly.(See Figure 4, and Figure 7).

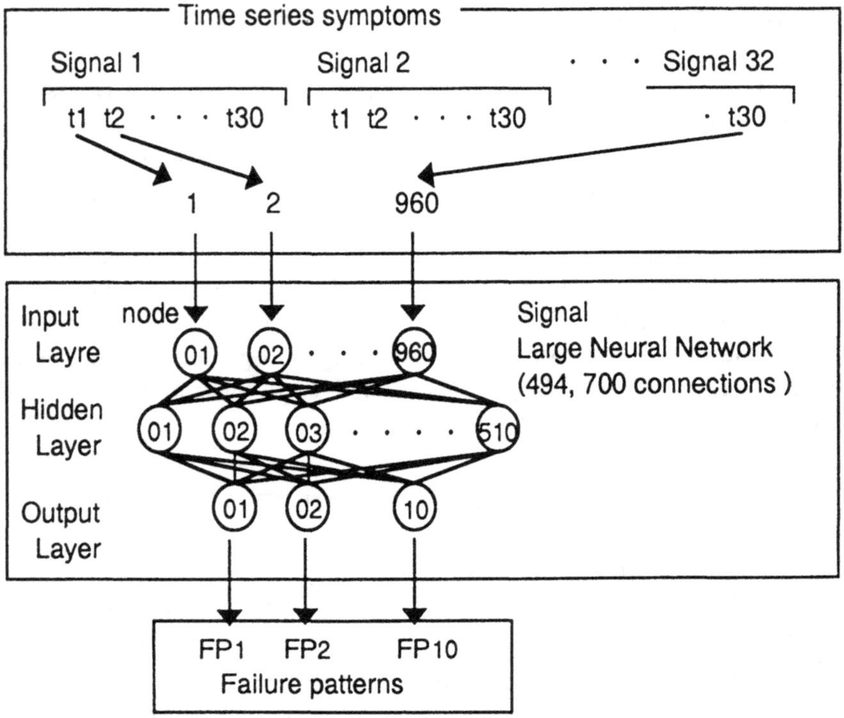

Figure 6 : Single large neural network system

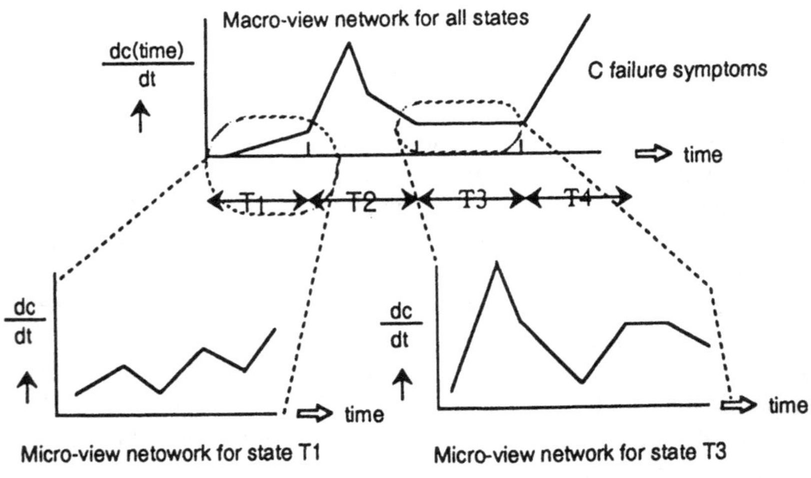

Figure 7 : Macro-view and micro-view for symptoms

When a very complex failure occurs in the ESS, the symptoms of four machine states are usually observed. Therefore, the four micro-viewing neural networks correspond to the four machine states. We added a module for integrating these distributed network employs an expert system. the reason for this is that: The output of a neural network includes ambiguous meaning. Then, when the input to the next neural networks uses such a output data, the next neural network output includes an unreliable judgment. Therefore, in the diagnostic problem domain, it is very difficult to join neural networks in series. The integration module includes information on the ESS machine architecture and the process of changing components. Using this knowledge, the system selects the final failure pattern from the reports of each distributed neural network. For Example, when a macro-viewing neural network is at a loss whether to classify the failure pattern as an N4T failure or an N2E failure, the integration module determines the final failure pattern(N4T) by using the opinions of each micro-viewing neural network. The configuration of the distributed neural networks is shown in Figure 8.

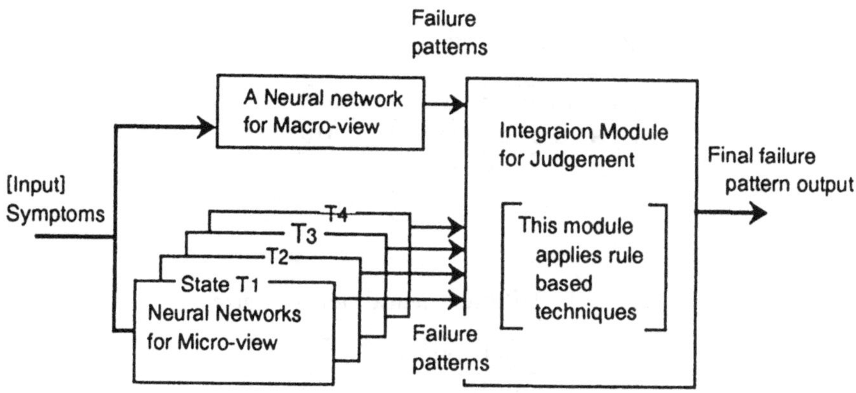

Figure 8 : Distributed neural network system

4 Evaluation.

The effectiveness of the distributed neural network system(Figure 8) is evaluated in terms of the failure pattern recognition rate. Table-1 compares the performance of this system with the traditional statistical method and a single neural network. Recognition was improved 20% by combination micro-viewing neural networks and a macro-viewing neural network.

The distributed system achieved close to 95% recognition in the training set (Table-1, Test-1), and 88% recognition with non-training data(Table-1, Test-2). The training data(100 failures) and non-training data(50 failures) were generated by a dummy fault generator in the ESS. The sizes of the single, large neural network and the distributed neural networks are shown in Table-2 and Table-3.

Table 1: Recognition rate

	Traditional Statistical method	Neural Network		
		Single Large-model	Only Macro-view Small-model	Distributed Macro-view Micro-view
Test-1	57.0%	85.9%	89.4%	95.2%
Test-2		58.0%	63.6%	88.6%

Test-1:Training Date Test-2:Non-Training Date

Table 2: Single,large neural network size

Input neurons	960
Hidden neurons	510
Output neurons	10
Connections	495×10^3

Table 3: Distributed neural network size

Each Network	
Input neurons	320
Hidden neurons	600
Output neurons	10
Connections	198×10^3

Here, the traditional statistical method refers to : (1) calculating differential time series data by time (dc/dt) for each signal, (2) summing these calculated values($\sum_{i=1}^{32} dc/dt$, i is signal number), (3) generating distribution patterns of $\sum_{i=1} dc/dt$, and (4) trying to match distribution patterns. Some failure patterns obtained by this traditional statistical method are shown in Figure 9 for reference.

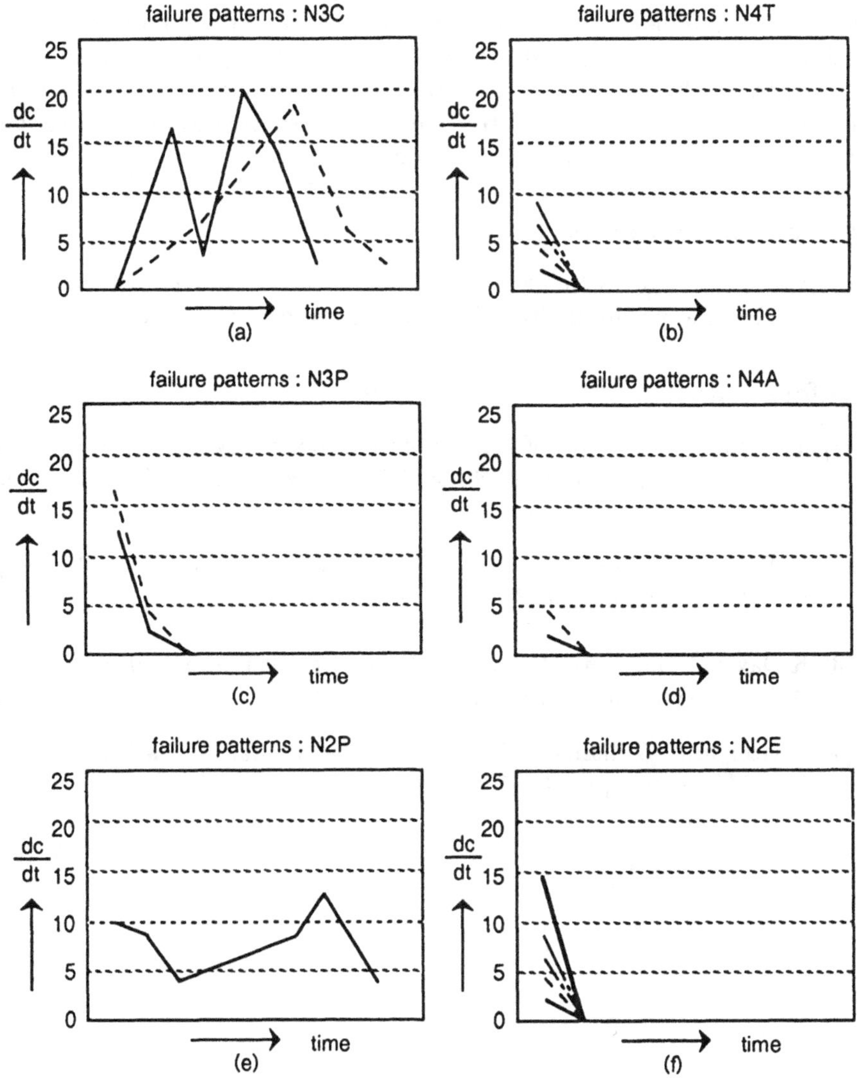

Figure 9 : Failure patterns by traditional statistical method

5 Conclusion

We have shown that the distributed neural networks system is very useful for diagnosis of complex intermittent faults in ESSs. This distributed neural network structure consists of four micro-viewing neural networks, one macro-viewing neural network and an integrated module of these distributed neural networks. We have evaluated the performance of the failure pattern recognition rate using real faults generated by the dummy fault generator. Using this distributed neural network system, the diagnostic rate was improved by 20.0% –30.0% compared with the traditional statistical method and a single neural network. Future studies are expected to focus on various improvements in the integration module.

Acknowledgements

Authors would like to thank M.Hirai, O.Saito, S.Yoshida, M.Inada, and F.Motoi for many helpful discussions during the preparation of this paper.

References

[1] T.Nogami, Y.Yokoi, M.Kasai, "Failure Diagnosis System on Pneumatic Control Valves by Neural Network," in *Proceedings ICNN.* 1993, p. 1876–1881.

[2] M.Boda, H.Brandt, "Application of Neural Networks in Fault Diagnose," in *Proceedings ISS,* vol. 1, October 1992, p. 254-257.

[3] R.Davis, "Diagnostic Reasoning Based on Structure and Behavior," *Artificial Intelligence,* vol. 24, 1984, p. 347–410.

[4] J.Kleer, B.C.Williams, "Diagnosis with behavioral Modes," in *Proceedings IJCAI,* 1989, p. 104–109.

A Neural Network Pre-processor for a Fault Diagnosis Expert System

Mr A. D. Chattell, Mr J. B. Brook

B67/210, B T Laboratories
Martlesham Heath, Ipswich. IP5 7RE. U K.
andyc@aom.bt.co.uk, jamesb@aom.bt.co.uk

Abstract

Line test data is important for identifying and locating faults in access networks. However, the complexity of the data makes devising suitable evaluation heuristics difficult. This paper describes the use of a neural network, trained using historical data, both as a stand-alone system and as a pre-processor for a rule-based expert system. Comparisons between the different cases are drawn and recommendations are made for the applicability of the different approaches.

1 Introduction

In common with other telecommunication operators' networks, the BT network comprises two main parts; the main trunk network and the access network. The trunk network is a highly modernised switched network using mainly optical fibre systems to route traffic between exchanges. This may be managed effectively, with most components being able to flag possible and actual faults using alarms and self test facilities. The access network provides the connection between the customers' premises and the nearest local exchange. For the provision of standard telephony service, this network predominantly comprises twisted copper pairs, one pair per telephone (un-switched). There are no active components between the network termination in the customer's premises and the telephone exchange, so testing the network must be managed from the exchange.

Each exchange houses one or more Line Test System (LTS) test heads. These are capable of making basic electrical parametric tests, such as measuring voltages on a given twisted copper pair. Each test head is connected to a LTS processor, which provides control and processes the test results. Each LTS processor is connected to the main customer service computer system.When a customer reports a fault on their telephone line, a receptionist can instigate a test on the customer's line using the LTS. The receptionist also symptomatically categorises the customer's description of the fault into one of a number of customer report codes (CRCs), e.g. NDT - no dial tone.

The location of the fault in the access network determines which engineering skill group should be despatched to repair it. There are three skill groups: customer apparatus (CA), local network (LN) and exchange (EX). Despatching the wrong person wastes time and hence money. Currently the despatch decision is made using an algorithm which uses both the parametric line test results and the fault category determined by the receptionist. If the results cannot be resolved to a particular despatch, the handling of the fault is passed to a Diagnostic Test Officer (DTO) who may find out other information relevant to the fault in order to determine a despatch.

There is considerable data available which could be useful in determining the correct despatch for a given fault, such as fault history data. This data could be used in addition to the line test data and the CRC in order to improve the accuracy of the despatch decision and hence reduce costs. An Expert System (ES) is being considered for using this data.

2 Fault Diagnosis Systems

2.1 Expert System Overview

An ES approach provides the flexibility to deal with different types of information. The fact that difficult faults are currently resolved by a human expert (the DTO) also indicates the suitability of using an ES.

An ES has been investigated which uses a Bayesian probabilistic rule-based approach. The main functional components of this system are shown in figure 1. The rule-base mainly comprises rules in the standard IF - THEN format. However, as the system is probabilistic, there are parameters associated with each rule which determine the effect of the rule. Data relating to the evaluation of each rule and the despatch decision made by the system are stored in the Decision Record. Once a fault has been fixed, the correct location and hence the correct despatch decision is known. The location at which a fault was found is

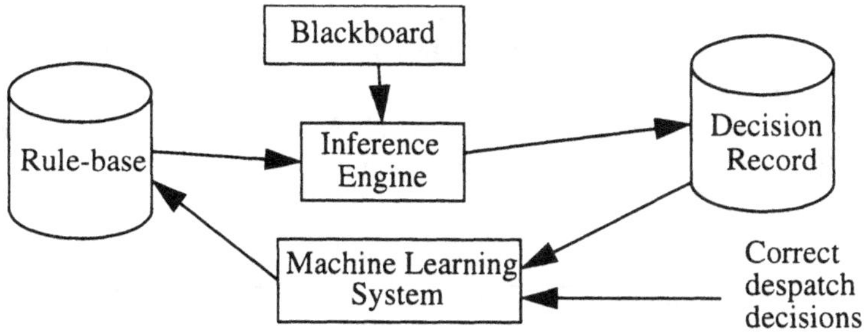

Figure 1. Main components of the Expert System.

represented numerically by a clear code. The correct location may be compared with the data stored in the Decision Record to allow the best weighting factors for each rule to be statistically calculated. The Machine Learning System (MLS) is responsible for this calculation. The system can therefore learn from its past decisions and improve its future despatch accuracy. The weighting factors for the rules also gives a direct quantified measure of the value of the rule. This indicates how important the rule may be in determining the final despatch decision.

The Inference Engine (IE) is responsible for the order of processing and the evaluation of the rules. If the rule uses an item of data which is unknown at the time of evaluation, the IE makes a request to the Blackboard to obtain the data. The Blackboard is the only real time interface with the outside world.

2.2 Acquisition of Rules for the ES

There are many ways of determining rules for use by an ES. The most commonly used, known as knowledge elicitation, is to interview human experts, in this case the DTOs. Humans are good at identifying qualitative relationships between real world events, e.g. "if the customer says this, then it is likely that the fault is with the customer's phone", but are poor at quantifying these relationships. They also have difficulty in determining relationships between more abstract events or data. The problem of quantifying relationships is taken care of by the Machine Learning System using past experience. This leaves the question of how to deal with more abstract data such as that produced by a LTS.

2.3 Using Neural Networks

This difficulty in interpreting LTS data prompted an investigation into the use of a Neural Network (NN) [3] [4] as a classifier to pre-process the data. The output from the NN could then be used by the ES. It was also decided to try the NN as a stand alone system. The various configurations for both the ES and NN are shown in figure 2.

The multi-layer perceptron (MLP) was the basic architecture chosen to implement the neural networks, owing to its ability to handle both the continuous and binary LTS data and because required outputs determined from the fault clear codes could be used for supervised training[1]. A highly configurable package, written at BT Labs, was available for implementing such networks.

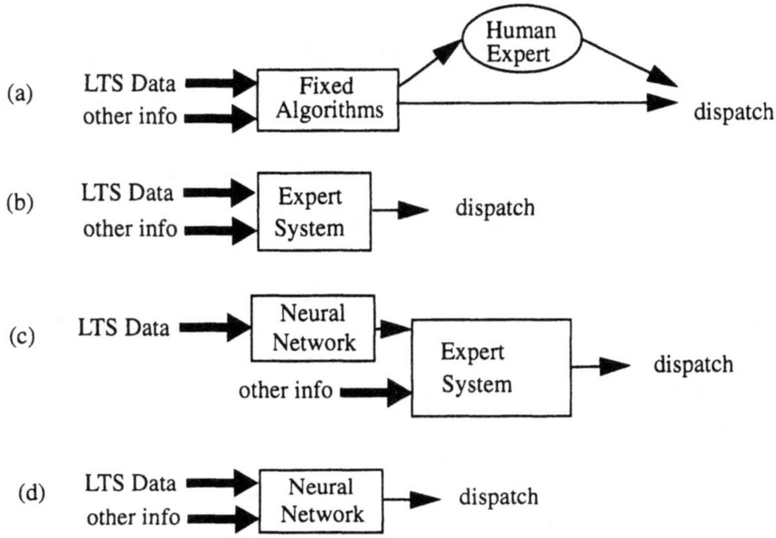

Figure 2. Fault diagnostic system architectures investigated

3 Using the MLP

3.1 Nature of the training data

The training data consisted of fault reports extracted from various BT databases. The data was obtained from over seventy exchanges, which between them used all four types of LTS used in BT's access network.

The line test results that were used by one or more networks in this investigation are listed below, where A and B refer to the two copper wires comprising a pair:

Voltages: A-Earth, B-Earth, A-B

Resistances: A-Battery, B-Battery, A-Earth, B-Earth, A-B, B-A

Capacitances: A-Earth, B-Earth, A-B

Termination (whether or not the LTS can recognise the customer premises termination components)

Some LTS types do not return all values for a particular fault. Of the 12,000 available faults, only 3,650 had all the required values. To use all 12,000, additional boolean *presence flag* inputs were provided. The smaller set was more typically used in experiments.

The plant item locations and corresponding despatch are shown in table 1. The

Table 1: Plant item locations.

Location	Mnem	Desp.
no fault	ok	OK
exchange	ex	EX
exchange/pcp connection	econn	LN
primary cross connection point	pcp	LN
exchange/dp or pcp/scp or pcp/dp	dconn	LN
secondary cross connection point	scp	LN
scp/dp connection	sconn	LN
distribution point	dp	CA
dropwire	dw	CA
customer apparatus	ca	CA

proportions of faults varies considerably across the locations.

3.2 Preparation of the training data

The data needed to be statistically processed in order to present it in a geometry suitable for training by back-propagation. In particular, some form of normalization was required, due to the different orders of magnitude of the various inputs. Specifically, a small relative change in resistance (order 10^5) could overwhelm an extreme relative change in capacitance (order 10^{-7}).

The method used for scaling was the Z-scoring process [2]. Using the mean (μ) and standard deviation (σ) of the data, the Z-score of a value X is calculated according to the following formula:

$$\text{Z-score} = (X - \mu) / \sigma$$

This produces a distribution of the data that is independent of its order of magnitude and units of measure. Ideally, both testing and training data should have the same values of μ and σ for this calculation. In practice, when testing on new faults, historical values should give a good approximation.

A further problem inherent in the resistance data was a very top-heavy distribution. Resistances are only measured up to 1 Mohm, yet the majority of values (over 95%) fell into this range.

For stand-alone MLPs (illustrated in fig. 2(d)), any enumerated data (such as the CRCs) must be represented in a binary form, with a boolean input for each possible value. This leads to a significant increase in the number of inputs.

This data was then split into training and testing sets. Training set sizes of 1000, 1500 or 2000 were used. It was found that the proportion of the training data for each class should be between the naturally occurring proportion and an equal split across all of the classes.

3.3 Network Topologies Used

A perceptron consists of a number of inputs, zero or more hidden layers and an

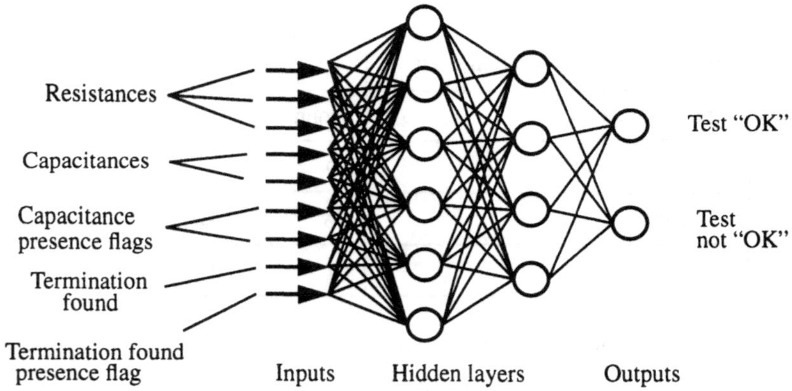

Figure 3. An example MLP topology.

output layer.

The number of inputs depended on the data set used, as illustrated in Figure 3. This example shows the use of three resistance values (always present), two capacitance values (not always present) and the line termination boolean (not always present).

The number of outputs depended on the problem to be solved, or more particularly the number of classes into which the input data was to be resolved. The early networks had 10 output classes, for the different possible plant items shown in Table 1. This was narrowed down to the dispatch decision, also shown in Table 1, then finally to two classes, representing the boolean "Test OK", i.e.

the network line is not faulty, (as shown in Figure 3). In addition, other topologies were tried, by combining various dispatch groups.

To put these topologies in the context of the ES, the first two (classifying according to location or dispatch) could in fact work as a stand alone system, bypassing the ES (Figure 2(d)), whereas the other MLPs could be used by rules within the ES (figure 2(c)).

The number of hidden layers used was found experimentally, and depended on the size of the training set. With each training set, various network topologies with either one or two hidden layers were tried, within the constraints of execution time. The number of nodes was increased on each execution, until the training and testing error percentages differed by more than 2%, showing a loss of generality of the MLP.

The networks used to obtain the results given in this paper typically used two hidden layers, with 12-16 nodes in the first hidden layer, and 3-4 in the second.

3.4 Training parameters

At each step during the back-propagation learning algorithm, the weight change was affected by the use of a momentum factor, the addition of random noise to the weights and the use of an adaptive version of the algorithm. These parameters helped to avoid local minima in the error function.

In addition, a bound was placed on the absolute value of the weights at each step. This was found to help produce more general networks, able to perform equally well with unseen testing data as with training data.

4 Results

These results are comparisons with the ES which has rules derived from existing despatch algorithms, not from any analysis of test data. The rules use the customer report code and the LTS results.

4.1 Using the MLP as a pre-processor

A version of the ES was created with an additional rule which alters fault location probabilities according to whether the LTS results were interpreted as "ok" or not "ok". The input for this rule came from an MLP with two output classes, representing the boolean "ok" value. The set of 3,650 faults was used for training and testing. Of these, 1500 values were used for training. The number of faults in the training set of the "ok" class was about equal to that of the "not ok" class. The network trained with 74% accuracy and tested with 71% accuracy on 2150 faults for the "ok"/"not ok" classification.

The overall despatch accuracy using the ES with the MLP pre-processing the LTS results was 4% higher than using the ES on its own.

4.2 Using the MLP on its own

Using just the LTS data as input and having output classes representing the possible despatch decisions, the performance of the MLP was about the same as for the ES. It was decided to try to use the customer report codes as well as the LTS data. Owing to the large number of CRCs, the fault data was analysed and the nine most frequently occurring values were chosen as inputs. The performance of the MLP increased to 11% higher than the ES on the same data set.

4.3 Enhancing the ES

An analysis of the fault data was made in order to identify rules. This may be considered to be analogous to the training of the MLP. This analysis was very manpower intensive. The performance of the ES was increased by 11% to virtually the same as that of the MLP.

5 Future work

The work done so far, identified several interesting points which could be looked into in the future.

A larger set of data is required, in order to gain more training sets for the less common fault locations. Alternatively, more training sets could be synthesised by adding noise to existing sets.

The network could be trained with the ability to recognise a situation where the LTS data can not be used for a diagnosis. This would be the equivalent of a DTO dispatch queue. Currently, only simple heuristics have been used to evaluate the output from the MLPs; the class returned is the one with the highest output. An alternative is to threshold the output and only return the highest class if its output exceeds the threshold, otherwise return "not sure enough".

More care needs to be taken to match the geometry of the inputs with the non-linear output function used. For example, if Z-scoring is being used, the data will be distributed equally about zero. Therefore boolean values should be represented as +/- 1, rather than as 0 and 1. Moreover, an appropriate sigmoidal non-linearity should be used, namely the hyperbolic tangent function with range +/- 1, rather than the current logistic function, which has range [0,1].

It could be informative to look at the weights and biases, to see which ones are strongly inhibitory or excitary. This could help determine ES rules.

It is better to use a larger number of classes, and regroup the subclasses during post-processing, rather than to use a smaller number of classes. This suggests that using MLPs for complete fault classification could have as much potential as an "OK"/not "OK" net. Furthermore, if the latter type of input (boolean) is still required for use by an ES, then certain outputs from a multi-class MLP could be grouped together by a post-processor.

6 Conclusions

Early experiments rapidly highlighted the need for a well prepared data set.

The main reason for choosing an Expert System approach for the fault location problem was the ease and flexibility with which it can use different types of data. The experiments carried out mainly used the same data as for existing algorithms, which largely comprises continuous real numbers. The MLP is particularly suited to data in this format. It required considerable work by someone knowledgeable about the LTS data to increase the performance of the ES to that of the neural network (MLP). Combining the MLP with the ES increased the performance of the ES without requiring the effort of an expert on the data.

The MLP performed as well as any system which used the ES, although for enumerated data, the number of inputs required very quickly becomes prohibitively large, and free format data cannot be used. Also, if new data becomes available, all previous training effort must be discarded.

There are three main conclusions to be drawn from this work. Firstly, the MLP is much better suited to processing continuous real numbers than the ES. However, the ES is much better suited to processing non-numeric data. Secondly, if there is an expert on the problem domain then an Expert System can be used, otherwise a neural net is likely to give better results. Thirdly, if the amount of available data is not static, i.e. it is likely to increase frequently in the future, then an ES is much more suitable than an MLP.

For a final fault diagnosis system, which could use a much wider variety of data than the experiments done so far, the strengths of both techniques would complement each other to give a better system than either could achieve alone.

References

[1] Lippmann, R.P. "An Introduction to Computing with Neural Nets", IEEE ASSP, April 1987, pp4-22.

[2] Crooks, T. "Care and Feeding of Neural Networks", AI Expert, July 1992, pp 36-41.

[3] Sethi, I.K. "Artificial Neural Networks and Statistical Pattern Recognition", North-Holland publishers.

[4] Dayhoff, J. "Neural Network Architectures", Van Nostrand Reinhold publishers.

Author Index

Subject Index

Q

R

S

T

V

W

www.ingramcontent.com/pod-product-compliance
Lightning Source LLC
LaVergne TN
LVHW011114300926
849160LV00008B/380
9781138882812